New Advances in the Research of Antioxidant Food Peptides

New Advances in the Research of Antioxidant Food Peptides

Editors

Lourdes Amigo Garrido
Blanca Hernández-Ledesma

MDPI • Basel • Beijing • Wuhan • Barcelona • Belgrade • Manchester • Tokyo • Cluj • Tianjin

Editors
Lourdes Amigo Garrido
Instituto de Investigación en
Ciencias de la Alimentación
(CIAL, CSIC-UAM)
Spain

Blanca Hernández-Ledesma
Instituto de Investigación en
Ciencias de la Alimentación
(CIAL, CSIC-UAM)
Spain

Editorial Office
MDPI
St. Alban-Anlage 66
4052 Basel, Switzerland

This is a reprint of articles from the Special Issue published online in the open access journal *Foods* (ISSN 2304-8158) (available at: https://www.mdpi.com/journal/foods/special_issues/New_Advances_Research_Antioxidant_Food_Peptides).

For citation purposes, cite each article independently as indicated on the article page online and as indicated below:

LastName, A.A.; LastName, B.B.; LastName, C.C. Article Title. *Journal Name* **Year**, *Volume Number*, Page Range.

ISBN 978-3-0365-2520-4 (Hbk)
ISBN 978-3-0365-2521-1 (PDF)

Contents

About the Editors

Lourdes Amigo Garrido (PhD)

Lourdes is a research professor at the Institute of Food Science Research (CIAL, CSIC-UAM), Madrid, Spain. She was director of the Institute of Industrial Fermentations (CSIC) (February 2002–July 2010), director of the Centre of Organic Chemistry "Manuel Lora-Tamayo" (June 2002–June 2004), and coordinator of the director's permanent commission of CSIC (September 2006–July 2010). She has participated in 66 different R+ D projects and published 120 research papers in journals and 9 book chapters. Prof. Amigo has presented 22 invited lectures and 106 oral communications and posters in international and national symposia. She is co-author of four patents. She has supervised 8 PhD theses and 12 master's theses. She is a teacher in different graduate and post-graduate specialization courses for graduates, technicians, and master's students in Spain and South America. She is a supervisor of graduate students and technical personnel. She is also a reviewer of 35 SCI journals and a research projects evaluator.

Blanca Hernández Ledesma (PhD)

Blanca Hernández Ledesma has been working at the Institute of Industrial Fermentations (CSIC, January 1999–December 2006), the Univ. of California at Berkeley (January 2007–November 2009), and the Institute of Food Science Research (CSIC-UAM, January 2010-current). Her research career has been focused on the study of the biological activity of food proteins/peptides, investigating their modifications during gastrointestinal digestion, their bioavailability, and their mechanisms of action, and developing processes to produce ingredients with improved techno-functional and organoleptic properties. She is author of 85 JCR articles and 38 book chapters (h-index 37). Her results have been presented in 64 International and 24 National Symposia. She is co-inventor of three patents. She has supervised 4 PhD theses and 17 master's theses. She has participated in 30 R+D Projects, and 3 contracts within the industry. She is member of the editorial committees of four books and nine journals, participating as guest editor of six Special Issues.

Editorial

Introduction to the Special Issue: New Advances in the Research of Antioxidant Food Peptides

Lourdes Amigo and Blanca Hernández-Ledesma *

Institute of Food Science Research (CIAL, CSIC-UAM, CEI UAM + CSIC), Nicolás Cabrera 9, 28049 Madrid, Spain; lourdes.amigo@csic.es
* Correspondence: b.hernandez@csic.es; Tel.: +34-91-0017970

Received: 1 December 2020; Accepted: 4 December 2020; Published: 7 December 2020

During cell metabolism, oxygen is partially reduced to reactive oxygen species (ROS) that play a physiological role in cellular processes, including proliferation, cell cycle and death, and signal transduction. However, ROS are responsible for attacking cell nucleophilic centers, causing lipid peroxidation, protein oxidation, and genetic alterations, including DNA damage, mutations, epigenetic changes, and genomic instability. Fortunately, the human body is equipped with an effective defense system to neutralize the toxicity of ROS. However, an imbalance provoked by either an over-production of ROS or a defect in the cellular defense system results in a state known as oxidative stress. This status and its subsequent damages to vital cellular components have been associated with numerous severe chronic disorders, such as cardiovascular and neurodegenerative diseases, diabetes, metabolic syndrome, intestinal inflammatory diseases, and cancer. In addition, oxidation reactions are responsible for food deterioration during processing and storage. In spite of their remarkable effectiveness, the endogenous antioxidant systems are not sufficient and humans are dependent on dietary antioxidants to maintain ROS concentrations at low levels. A number of natural antioxidants have been revealed as potential preventative/therapeutic agents against oxidative stress. Among them, peptides from animal and vegetal food sources have attracted attention because of the plentiful evidence of their in vitro antioxidant properties. In addition to their potential as safer alternatives to synthetic antioxidants used to prevent oxidative reactions in foods, antioxidant peptides can also act by reducing the risk of numerous oxidative stress-associated diseases. Furthermore, peptides can act synergistically with non-peptide antioxidants, enhancing their protective effect.

This Special Issue of the *Foods* journal includes seven outstanding papers describing examples of the most recent advances in the antioxidant peptide research. It begins with a group of papers describing the antioxidant potential of vegetal food-derived hydrolyzates and peptides. The review of Olivares-Galván et al. [1] summarized the existing evidence on antioxidant peptides released from fruit residues, focusing on the current techniques used in their extraction, purification, fractionation, and identification, the strategies followed to allow the peptides' release from source protein, as well as the assays used to determine their antioxidant activity. Two research articles described the potential of two vegetal proteins as sources of antioxidant peptides. Thus, the study of Kusumah et al. [2] aimed at identifying the mung bean (Vigna radiata) albumin sequences with antioxidant capacity by both in silico and in vitro assays. The hydrolysates produced by thermolysin showed high antioxidant potential in terms of ferrous ion chelating and ORAC values because of the high hydrophobicity and low molecular mass of the released peptides. In the study of Esfandi et al. [3], hydrolyzates from oat bran protein with Alcalase®, Flavourzyme®, papain or Protamex® showed the ability to protect hepatic HepG2 cells against oxidative damage by reducing ROS levels and caspase-3 activity, and increased glutathione concentration and antioxidant enzymes activity.

The following articles focused on animal protein sources of antioxidant peptides. First, the review of Gilmartin et al. [4] described the potential role exerted by whey protein, hydrolyzates or peptides in the modulation of sarcopenic biomarkers in myoblast cell lines, and in aged rodents and humans.

The human intervention trials have shown that a daily dietary supplementation of 35 g of whey is likely to improve sarcopenic biomarkers, improving muscle mTOR signaling in the elderly, although exercise appears to have the greatest benefit for old muscle. Kleekayai et al. [5] hydrolyzed bovine whey protein concentrate with Debitrase™ and FlavorPro™ under pH-stat and non pH-controlled conditions, evaluating the impact of hydrolysis conditions on the physicochemical and antioxidant activities of the released hydrolyzates. These authors demonstrated that hydrolyzates generated under pH-stat conditions displayed higher radical scavenging activities than those shown by non-pH-controlled conditions. Moreover, a more pronounced impact of conditions in the biochemical assays compared to the cellular antioxidant assay was observed. In the study of Amigo et al. [6], an integrated approach combining in silico and in vitro assays was used, confirming the multifunctionality of milk and meat protein-derived peptides that were similar to or shared amino acids with previously described opioid peptides. Fifteen of twenty-seven assayed peptides were found to exert two or more activities, with angiotensin-converting enzyme (ACE) inhibitory, antioxidant, and opioid being the most commonly found. Four fragments, RYLGYLE, YLGYLE, YFYPEL, and YPWT, demonstrated ACE-inhibitory and antioxidant activities, and also protected Caco-2 and macrophage RAW264.7 cells against chemical-induced oxidative damage. Finally, the review of Benedé and Molina [7] summarized the current knowledge on the antioxidant activity of chicken egg proteins and their derived peptides. The main process for obtaining these bioactive peptides from their source proteins is enzymatic hydrolysis using enzymes and/or processing technologies, such as heating, sonication or a high-intensity-pulsed electric field. Different in vitro assays have been used to evaluate the mechanisms of action of egg bioactive peptides, the in vivo effects of which have been confirmed by both cell culture assays and animal models.

Author Contributions: Writing—original draft preparation, L.A., and B.H.-L.; writing-review and editing, L.A., and B.H.-L.; funding acquisition, L.A., and B.H.-L. All authors have read and agreed to the published version of the manuscript.

Funding: This work was funded by project AGL2015-66886-R from the Spanish Ministry of Science and Innovation (MICIU) and PID2019-104218RB-I00 (CSIC).

Acknowledgments: We wish to thank the invited authors for their interesting and insightful contributions.

Conflicts of Interest: The authors declare no conflict of interest.

References

1. Olivares-Galván, S.; Marina, M.L.; García, M.C. Extraction and Characterization of Antioxidant Peptides from Fruit Residues. *Foods* **2020**, *9*, 1018. [CrossRef] [PubMed]
2. Kusumah, J.; Real Hernandez, L.M.; Gonzalez de Mejia, E. Antioxidant Potential of Mung Bean (*Vigna radiata*) Albumin Peptides Produced by Enzymatic Hydrolysis Analyzed by Biochemical and In Silico Methods. *Foods* **2020**, *9*, 1241. [CrossRef] [PubMed]
3. Esfandi, R.; Willmore, W.G.; Tsopmo, A. Antioxidant and Anti-Apoptotic Properties of Oat Bran Protein Hydrolysates in Stressed Hepatic Cells. *Foods* **2019**, *8*, 160. [CrossRef] [PubMed]
4. Gilmartin, S.; O'Brien, N.; Giblin, L. Whey for Sarcopenia; Can Whey Peptides, Hydrolysates or Proteins Play a Beneficial Role? *Foods* **2020**, *9*, 750. [CrossRef] [PubMed]
5. Kleekayai, T.; Le Gouic, A.V.; Deracinois, B.; Cudennec, B.; FitzGerald, R.J. In Vitro Characterisation of the Antioxidative Properties of Whey Protein Hydrolysates Generated under pH- and Non pH-Controlled Conditions. *Foods* **2020**, *9*, 582. [CrossRef] [PubMed]
6. Amigo, L.; Martínez-Maqueda, D.; Hernández-Ledesma, B. In Silico and In Vitro Analysis of Multifunctionality of Animal Food-Derived Peptides. *Foods* **2020**, *9*, 991. [CrossRef] [PubMed]

7.   Benedé, S.; Molina, E. Chicken Egg Proteins and Derived Peptides with Antioxidant Properties. *Foods* **2020**, *9*, 735. [CrossRef] [PubMed]

Publisher's Note: MDPI stays neutral with regard to jurisdictional claims in published maps and institutional affiliations.

Article

Antioxidant Potential of Mung Bean (*Vigna radiata*) Albumin Peptides Produced by Enzymatic Hydrolysis Analyzed by Biochemical and In Silico Methods

Jennifer Kusumah, Luis M. Real Hernandez and Elvira Gonzalez de Mejia *

Department of Food Science and Human Nutrition, University of Illinois at Urbana-Champaign, Urbana, IL 61801, USA; kusumah2@illinois.edu (J.K.); rhealmlus@gmail.com (L.M.R.H.)
* Correspondence: edemejia@illinois.edu; Tel.: +1-217-244-3196; Fax: +1-217-265-0925

Received: 13 July 2020; Accepted: 2 September 2020; Published: 4 September 2020

Abstract: The objective of this study was to investigate the biochemical antioxidant potential of peptides derived from enzymatically hydrolyzed mung bean (*Vigna radiata)* albumins using an 2,2′-Azino-bis(3-ethylbenzothiazoline-6-sulfonic acid) (ABTS) radical scavenging assay, a ferrous ion chelating assay and an oxygen radical absorbance capacity (ORAC) assay. Peeled raw mung bean was ground into flour and mixed with buffer (pH 8.3, 1:20 *w/v* ratio) before being stirred, then filtered using 3 kDa and 30 kDa molecular weight cut-off (MWCO) centrifugal filters to obtain albumin fraction. The albumin fraction then underwent enzymatic hydrolysis using either gastrointestinal enzymes (pepsin and pancreatin) or thermolysin. Peptides in the hydrolysates were sequenced. The peptides showed low ABTS radical-scavenging activity (90–100 µg ascorbic acid equivalent/mL) but high ferrous ion chelating activity (1400–1500 µg EDTA equivalent/mL) and ORAC values (>120 µM Trolox equivalent). The ferrous ion chelating activity was enzyme- and hydrolysis time-dependent. For thermolysin hydrolysis, there was a drastic increase in ferrous ion chelating activity from t = 0 (886.9 µg EDTA equivalent/mL) to t = 5 min (1559.1 µg EDTA equivalent/mL) before plateauing. For pepsin–pancreatin hydrolysis, there was a drastic decrease from t = 0 (878.3 µg EDTA equivalent/mL) to t = 15 (138.0 µg EDTA equivalent/mL) after pepsin was added, but this increased from t = 0 (131.1 µg EDTA equivalent/mL) to t = 15 (1439.2 µg EDTA equivalent/mL) after pancreatin was added. There was no significant change in ABTS radical scavenging activity or ORAC values throughout different hydrolysis times for either the thermolysin or pepsin–pancreatin hydrolysis. Overall, mung bean hydrolysates produced peptides with high potential antioxidant capacity, being particularly effective ferrous ion chelators. Other antioxidant assays that use cellular lines should be performed to measure antioxidant capacity before animal and human studies.

Keywords: albumin; albumin peptide; antioxidant peptide; bioactive peptide; in silico; mung bean; mung bean albumin; peptide sequencing; *Vigna radiata*

---

1. Introduction

Mung bean, also known as green gram, is a small, green-colored legume widely cultivated throughout Asia [1]. It is a popular legume in countries such as Indonesia and China where its consumption is associated with positive health outcomes [2,3]. Mung bean flour is commonly made into a paste and incorporated into bread and desserts [4]. Mung beans have a relatively high protein content (19.5–33.1%) that is comparable to that of soybeans (*Glycine max*) (35–50%) and kidney beans (*Phaseolus vulgaris*) (23–25%) [5–7]. Due to its nutritional content, mung bean can be a plant-based protein source in developing nations where animal protein sources are cost-prohibited [8]. Compared to other legumes, mung beans are relatively free from antinutritional factors [9]. Mung beans are also rich in vitamins and minerals such as iron, magnesium, potassium, copper and folate [10].

The major storage proteins of mung bean are globulins (62.0%), albumins (16.3%), glutelins (13.3%) and prolamins (0.9%), with vicilin-type protein (8S) making up 89% of globulins [11]. Globulins are the main storage proteins in mature mung beans, and they are also the most well studied mung bean proteins [12]. In contrast to globulins, there are limited studies on the albumin proteins of mung beans [13]. Albumins are water-soluble, globular proteins found in both animals and plants [14]. In plants, albumins can be proteins stored in seeds to be used during germination and growth [15]. The isolation and characterization of some albumins in multiple legumes, such as lentils, soybeans and winged beans, has already been performed [16–18]. There are currently two main entries for mung bean albumin in the UniProt Database; Q9FRT8 is a reviewed entry detailing a 10 kDa protein fragment. Q43680 is a non-reviewed entry detailing a 30 kDa protein. The sequences of these two entries are given in Figure 1, and the sequences have minimal similarity (7% identity). There is currently no evidence that only a single mung bean albumin exists, so reported mung bean albumin sequences can be different from each other. Given the current data, mung bean albumins are not expected to have a molecular weight greater than 30 kDa [19,20].

```
Q9FRT8 ALB1_VIGRR        1    -------------------------------------------------AADCNGACSP
Q43680 Q43680_VIGRR      1    MSNLPYINAAFRFSSRDYEVYFFAKNKYVRLQYTPGKTEDKILTNLRLISSGFPSLAGTP
                                                                            :.   .  *  :*

Q9FRT8 ALB1_VIGRR       11    FEMPP-------------------------------------------------------
Q43680 Q43680_VIGRR     61    FAEPGIDSAFHTEASEAYVFSANNRAYIDYAPGTTNDKILAGPTTIAEMFPVLRNTVFAD
                              *   *

Q9FRT8 ALB1_VIGRR       16    -----------------------------------------------------------
Q43680 Q43680_VIGRR    121    SIDSAFRSTKGKEVYLFKGNKYVRIDYDSKQLVGSIRNISDGFPVLNGTGFESGIDASFA

Q9FRT8 ALB1_VIGRR       16    -------------------------CRSTDC------RCI--PIALFGGFCINPTGLSSVAKMI
Q43680 Q43680_VIGRR    181    SHKEPEAYLFKGDKYVRIHFTPGKTDDTLVGDVRPILDGWPVLKAFCLCELNKPSLSCI-
                                                      ::  *          *  *      ::  .**:    .   *:: :

Q9FRT8 ALB1_VIGRR       47    DEHPNLCQSDDECLKKGSGNFCARYPNHYMDYGWCFDSDSEAL
Q43680 Q43680_VIGRR    240    -NHLSLVTI----NKAFISNVCLFFFNVTLGLEACFLS-----
                              :* .*        *    .*.*  : *  :.    ** *
```

Figure 1. Alignment of the mung bean albumin sequences currently present in the UniProt Database. The alignment was conducted using Clustal Omega. Q9FRT8 is the reviewed sequence of a 10 kDa protein fragment, while Q43680 is the non-reviewed sequence of a 30 kDa protein. The sequences have 7% identity, with 19 identical positions and 21 similar positions. Asterisks (*) indicate identical amino acids in the sequences, while dots indicate amino acids in the sequences that are not identical.

Peptides derived from chickpea albumins, which are legume albumins like those present in mung beans, have shown high biologically relevant antioxidant potential [21,22]. Antioxidant peptides can benefit human health by chelating excess transition metal ions and scavenging free radicals and reactive oxygen species [23]. Lunasin, a peptide derived from soybean 2S albumin, has also been found to have antioxidant effects [24,25]. Peptides from whole mung bean protein hydrolysates have been found to have calcium and ferrous ion binding activity that can have biological implications, but this activity is also useful in preventing oxidation in food systems [26]. However, the antioxidant potential of mung bean albumin hydrolysates and peptides alone, without the presence of other mung bean proteins, has been poorly studied [27]. The objective for this study was to investigate the biochemical antioxidant potential of peptides derived from enzymatically hydrolyzed mung bean albumins using either thermolysin or gastrointestinal enzymes pepsin and pancreatin, followed by sequencing and characterizing the peptides. Antioxidant potential was investigated using an ABTS radical scavenging assay, a ferrous ion chelating assay and ORAC assay.

2. Materials and Methods

2.1. Materials

Peeled and split raw mung bean dry seeds were purchased locally (Asian Best, Thailand), and stored at 4 °C. The purchased mung beans were from a single brand cultivated in Thailand. The beans were taken from 4 °C bulk storage for each experimental sampling, at least three times for each experiment performed.

Centrifugal ultrafiltration filters with 3 or 30 kDa MWCO membranes were purchased from Millipore-Sigma (St. Louis, MO, USA). Protein reagents A and B, 2× Laemmli sample buffer, 10× tris/glycine and 10× tris/glycine/SDS buffers, mini-PROTEAN® TGX™ gels, Coomassie blue and Precision Plus Protein™ Dual Xtra standard were purchased from Bio-Rad (Hercules, CA, USA).

Simply Blue Safe Stain was purchased from Invitrogen (Carlsbad, CA, USA). All other reagents were purchased from Sigma-Aldrich (St. Louis, MO, USA) unless stated otherwise.

2.2. Mung Bean Albumin Extraction

Peeled dry mung beans were milled and sieved using a number 35 mesh (500 μm). Collected flour was stored at 4 °C until use. Since a standard extraction protocol for mung bean albumins has not been established, two different legume albumin extraction methods were implemented to extract mung bean albumins.

For one method, a recently proposed protocol for the extraction of mung bean albumins was followed as specified in Du et al. [28] with specific modifications. The collected mung bean flour was mixed with double-distilled water at a ratio of 1:20 (*w/v*). The pH of the mixture was adjusted to either 3.0 or 7.5 with 1 M HCl or NaOH, respectively. The extraction pH values used were previously found to solubilize a higher concentration of mung bean albumins compared to total mung bean proteins [28]. Albumins were extracted for 1 h at 25 °C with manual stirring of each mixture every 15 min. The mixtures were then centrifuged at 11,000× *g* for 20 min at 4 °C, and the collected supernatants had their pH adjusted to 4.6 using 0.1 M HCl or NaOH. The supernatants at pH 4.6 were centrifuged at 11,000× *g* for 20 min at 4 °C, and the precipitated mung bean albumin pellets were collected. The mung bean albumin pellets for each extraction pH were solubilized individually in 5 mL of double-distilled water (pH = 11) and sonicated for 1 min to facilitate solvation. The resulting solutions were filtered using 30 kDa MWCO ultrafiltration centrifugal filters by centrifuging the solutions at 6600× *g* for 1 h at 4 °C. The resulting permeate was collected and used as an aqueous filtrate of extracted mung bean albumins. The collected permeates were stored at 4 °C for ≤3 days.

The other extraction method followed the protocol published by Singh, Rao and Singh [29] with minor adaptations. Collected mung bean flour was mixed with 0.1 M borate buffer (pH = 8.3) at a ratio of 1:20 (*w/v*) and was continuously stirred for 1 h. The mixture was then centrifuged at 11,000× *g* for 20 min at 4 °C, and the supernatant was collected. As an alternative to extensive dialysis, the collected supernatant was filtered using a 3 kDa MWCO ultrafiltration centrifugal filter at 6600× *g* for 1 h at 4 °C, and then the retentate was supplied with sodium citrate buffer (pH = 4.6) to replenish the buffer volume lost as permeate. The solution at pH 4.6 was centrifuged at 11,000× *g* for 20 min at 4 °C, and the collected supernatant was filtered using a 30 kDa MWCO ultrafiltration centrifugal filter at 6600× *g* for 1 h at 4 °C. The resulting permeate was collected and used as an aqueous filtrate of extracted mung bean albumins. The collected permeates were stored at 4 °C for ≤3 days

2.3. Gel Electrophoresis

2.3.1. Sodium Dodecyl Sulfate–Polyacrylamide Gel Electrophoresis (SDS-PAGE)

The DC protein assay was used to determine the protein concentrations of mung bean albumin samples using a bovine serum albumin standard curve. SDS-PAGE was performed as published before with minor modifications [30]. Briefly, mung bean samples were mixed with 2× Laemmli buffer

(1:1 *v/v*) without 2-mercaptoethanol and boiled for 5 min. The prepared samples were loaded on 4–20% Tris-glycine gels so that each well had 20 μg of crude mung bean albumins (unfiltered extract) or 15 μg of filtered mung bean albumins (≤30 kDa permeate). A Tris-glycine buffer consisting of 25 mM Tris and 192 mM glycine at pH 8.3 was used in both the anode and cathode, and the gels ran at 200 V for 30 min. SimplyBlue Safe Stain was used to stain the gels.

2.3.2. Native Blue–Polyacrylamide Gel Electrophoresis (Native Blue-PAGE)

The SDS-PAGE methodology mentioned above was modified to run native blue gels. Mung bean samples were mixed (1:1 *v/v*) with Tris-glycine buffer containing 25% glycerol. Prepared samples were loaded on 4–20% Tris-glycine gels so that each well had 20 μg of crude mung bean albumins (unfiltered extract) or 15 μg of filtered mung bean albumins (≤30 kDa permeate). Tris-glycine buffer was used in the anode, and Tris-glycine buffer containing 2% Coomassie blue dye was used in the cathode. The gels ran at 150 V for 1 h at 22 °C. The gels were de-stained with a 25% isopropanol solution (*v/v*) containing 10% acetic acid (*v/v*).

2.4. In Silico Hydrolysis of Mung Bean Albumin Sequences

To determine proteases, outside of those present during gastrointestinal digestion, capable of potentially producing antioxidant peptides from mung bean albumins, a manual theoretical hydrolysis of the two reported mung bean albumin sequences (Q9FRT8 and Q43680) present in the UniProt Database (https://www.uniprot.org/) was performed. The theoretical hydrolysis was carried out using the protease specificity data present in the MEROPS Database (https://www.ebi.ac.uk/merops/). Alcalase, stem bromelain, ficin, papain and thermolysin were selected for the theoretical hydrolysis as they are food-safe enzymes capable of being used in the commercial hydrolysis of mung bean albumins.

For each mung bean albumin sequence, fragments of ≤8 adjacent amino acids, corresponding to an amino acid sequence that could occupy sites P4-P4′ in the active site of a protease, were matched to the possible amino acids known to be present at the P4-P4′ sites of a protease when a protein substrate is hydrolyzed by that protease. The specificity of each protease analyzed is detailed in Table 1. If a fragment sequence matched the possible amino acids that would lead the sequence to be hydrolyzed by a specific protease, the corresponding P1 and P1′ amino acids in the fragment sequence were color-coded, as this is where the sequence would be expected to be hydrolyzed. Amino acids in the mung bean albumin sequences were color-coded either red or purple where hydrolysis would be expected to happen, and hydrolysis was expected to be possible between adjacent amino acids in the sequence that were color-coded the same color. Possible mung bean albumin peptide sequences that would be expected to be produced by each protease were obtained by cutting each mung bean sequence between two adjacent amino acids that were color-coded the same to produce fragments of various amino acid quantities. For conciseness and applicability, only possible di- and tri-peptides that could be produced for each protease were accounted for, as these small peptides have intestinal transporters that make them more bioavailable than their larger counterparts [31]. The bioactive fragments present in the two mung bean albumin sequences hydrolyzed were identified using the Database of Bioactive Peptides—BIOPEP Database (http://www.uwm.edu.pl/biochemia/index.php/pl/biopep).

Table 1. Possible Amino Acids at Positions in or Around Cleavage Site for Specific Proteases.

Protease	P4 Position	P3 Position	P2 Position	P1 Position	P1′ Position	P2′ Position	P3′ Position	P4′ Position
Alcalase	Gly, Pro, Ala, Val, Leu, Met, Phe, Tyr, Ser, Thr, Asn, Glu, His	Gly, Ala, Val, Leu, Phe, Ser, Thr, Asn, Gln, Asp, Glu, Lys, Arg, His	Gly, Pro, Ala, Val, Leu, Ile, Phe, Tyr, Ser, Thr, Asn, Gln, Glu, Arg, His	Ala, Val, Leu, Met, Phe, Tyr, Ser, Asn, Gln, Glu, Lys, Arg, His	Gly. Ala, Val, Leu, Met, Phe, Tyr, Ser, Thr, Gln, His	Gly, Pro, Ala, Val, Leu, Ile, Phe, Tyr, Ser, Thr, Gln, Glu, Arg, His	Gly, Pro, Ala, Val, Leu, Met, Phe, Tyr, Ser, Thr, Asn, Lys, Arg, His	Gly, Pro, Ala, Val, Leu, Tyr, Ser, Thr, Cys, Asn, Gln, Asp, Glu, Lys, His
Stem Bromelain	Pro, Ala, Leu, Ile, Phe, Tyr, Ser, Thr, Cys, Glu, Arg, His	Gly, Pro, Ala, Val, Leu, Phe, Ser, Thr, Lys, Arg, His	Gly, Pro, Val, Leu, Phe, Tyr, Ser, Thr, Asn, Arg, His	Gly, Ala, Val, Leu, Phe, Tyr, Ser, Thr, Asn, Gln, Arg	Gly, Val, Leu, Ile, Phe, Tyr, Ser, Thr, Asn, Gln, Asp, Glu, Lys, His	Pro, Ala, Val, Leu, Tyr, Ser, Thr, Asn, Gln, Asp, Glu, Lys, His	Gly, Pro, Val, Leu, Phe, Ser, Thr, Cys, Asn, Gln, Asp, Glu, Arg, His	Gly, Pro, Ala, Val, Leu, Phe, Tyr, Ser, Thr, Asp, Glu, Lys, Arg, His
Ficin	Pro, Leu, Ser, Glu, Lys, Arg, His	Gly, Ala, Val, Leu, Ile, Phe, Thr, Arg, His	Gly. Val, Leu, Phe, Thr, Lys	Gly, Leu, Phe, Tyr, Ser, Lys, Arg, His	Leu, Phe, Tyr, Ser, Thr, Lys, His	Val, Leu, Tyr, Thr, Asn, Lys, His	Pro, Ala, Val, Leu, Ser, Thr, Glu, Lys	Gly, Pro, Val, Asn, Asp, Glu, Lys
Papain	Gly, Pro, Ala, Val, Leu, Ile, Phe, Tyr, Ser, Thr, Cys, Asn, Asp, Glu, Arg, His	Gly, Pro, Ala, Val, Leu, Ile, Met, Phe, Tyr, Ser, Gln, Asp, Glu, Lys, Arg, His	All 20 Canonical AA	All 20 Canonical AA	Gly, Ala, Val, Leu, Ile, Met, Phe, Tyr, Ser, Thr, Asn, Gln, Asp, Glu, Lys, His	Gly, Pro, Ala, Val, Leu, Ile, Phe, Tyr, Ser, Thr, Cys, Asn, Gln, Glu, Arg, His	Gly, Pro, Ala, Val, Leu, Phe, Tyr, Ser, Cys, Glu, Lys, Arg, His	Gly, Pro, Ala, Val, Leu, Ile, Phe, Tyr, Ser, Thr, Asn, Asp, Glu, Lys, Arg, His
Thermolysin	All 20 Canonical AA	All 20 Canonical AA	All 20 Canonical AA	All 20 Canonical AA	Gly, Pro, Ala, Val, Leu, Ile, Met, Phe, Tyr, Trp, Ser, Thr, Asn, Gln, Asp, Glu, Lys, His	All 20 Canonical AA	All 20 Canonical AA	All 20 Canonical AA

AA = Amino Acids.

2.5. Enzymatic Hydrolysis

The simulated gastrointestinal digestion of mung bean albumins was carried out using conditions used previously by Mojica and de Mejia [32]. Briefly, pepsin was added to a solution (pH = 2) of mung bean albumins (≤30 kDa fraction) at a ratio of 1:20 *w/w* for 2 h at 37 °C. After 2 h, pancreatin was added to the solution at a pepsin–mung bean albumin ratio of 1:20 *w/w*. The pH of the solution was increased to 7.5 using 1 M NaOH, and the solution was incubated for another 2 h at 37 °C. Enzymatic reactions were stopped by heating the solution to 90 °C for 15 min. The inactivated hydrolysate was stored at 4 °C for ≤2 days.

Thermolysin hydrolysis was carried out by adding thermolysin to a mung bean albumin solution (≤30 kDa fraction) at a thermolysin–mung bean albumin ratio of 1:20 *w/w*. The pH of the solution was adjusted to 8.0 using 1 M NaOH, and then the mixture was incubated at 70 °C for 4 h. Afterwards, the solution was heated to 95 °C for 15 min to stop proteolytic activity. The inactivated hydrolysate was stored at 4 °C for ≤2 days.

2.6. Antioxidant Assays

All antioxidant assays were carried out on fresh hydrolysates produced within ≤2 days of storage before the assays were conducted. Radical scavenging and ferrous ion chelating activities were calculated based on equations obtained from the standard curves using ascorbic acid and EDTA solutions, respectively.

2.6.1. ABTS Radical Scavenging Assay

The ABTS radical scavenging assay was performed as published before for a mung bean meal hydrolysate [26]. A 7 mM ABTS and 2.5 mM potassium persulphate stock solution in 10 mM PBS buffer (pH = 7.4) was made and stored in darkness for 16 h. The stock solution was then diluted with 10 mM PBS buffer until an absorbance of 0.70 ± 0.05 at 734 nm was obtained. A sample of mung bean albumins (20 μL, 1 g/L) was added to 1980 μL of diluted ABTS solution. The mung bean albumin concentration was determined by the DC protein assay using a bovine serum albumin (BSA) standard curve, and preliminary studies were performed to determine the concentration used. The reaction was allowed to react for 5 min in the dark, and the resulting absorbance was read at 734 nm. The results were expressed as ascorbic acid equivalent in μM.

2.6.2. Ferrous Ion Chelation Assay

The chelating activity of ferrous ions by mung bean peptides was analyzed as published before for a mung bean meal hydrolysate [26]. Mung bean albumins (100 μL, 1 g/L) were mixed with 100 μL of a 2 mM iron chloride solution, and the mixture was diluted with 1400 μL of double distilled water. Mung bean albumin concentration was determined by the DC protein assay using a bovine serum albumin (BSA) standard curve, and preliminary studies were performed to determine the concentration used. The mixture was incubated for 3 min at 25 °C. Afterwards, 400 μL of a 5 mM ferrozine solution was added, and the solution was incubated for 10 min at 25 °C. EDTA was used as a standard. The resulting absorbance was read at 562 nm. The result was expressed as EDTA equivalent in μM.

2.6.3. Oxygen Radical Absorbance Capacity (ORAC) Assay

An ORAC assay on mung bean albumin hydrolysates was performed with some modification as published before [33]. Briefly, 20 μL of the sample (1 g/L) was mixed with 120 μL of fluorescein (0.12 mM). The absorbance was then read at 485 nm and the mixture was incubated at 37 °C for 15 min. A total of 60 μL of 40 mM 2′2-Azobis(2-amidinopropane) dihydrochloride (AAPH) was then added, and the solution was read again at 582 nm. Trolox was used as standard. The results were calculated using Equations (1) and (2) and reported as Trolox equivalent (µM):

$$AUC = \left(\frac{R_1}{R_1}\right) + \left(\frac{R_2}{R_1}\right) + \left(\frac{R_3}{R_1}\right) + \ldots + \left(\frac{R_n}{R_1}\right) \tag{1}$$

$$Net\ AUC = AUC_{sample} - AUC_{blank} \tag{2}$$

2.7. Peptide Sequencing

Peptide sequencing was performed as published before [34]. Briefly, peptides obtained from the different hydrolysates were analyzed by HPLC–ESI–MS/MS using a Q-ToF Ultima mass spectrometer (Waters, Milford, MA, USA) equipped with an Alliance 2795 HPLC system. The gradient mobile phase A was 95% water, 5% acetonitrile and 0.01% formic acid, while mobile phase B was 95% acetonitrile, 5% water and 0.1% formic acid. The volume of injection was 400 µL/min and the PDA detector wavelength was 280 nm. Each peak was analyzed in MassLynx V4.1 software (Waters Corp., Milford, MA, USA) and the sequence of amino acids was identified based on the accurate mass measurements, while tandem MS fragmentation using the MassBank database was used to analyze the data and obtain the peptide sequences with >80% of certainty. The isoelectric point, net charge and hydrophobicity of the peptides were analyzed by PepDraw [35]. The amino acids were presented as one letter nomenclature.

2.8. Statistical Analysis

The experiments were repeated at least three independent times from different beans, starting from the mung bean albumin extraction and proceeding up to the antioxidant assays and peptide sequencing. Data are expressed as mean ± standard deviation. The data obtained were analyzed using one-way ANOVA to compare experimental to control values, and differences were considered significant at $p < 0.05$. GraphPad Prism 7 (GraphPad Software, LLC., San Diego, CA, USA) was used for the data analysis.

3. Results

3.1. Mung Bean Albumin Profiles

The mung bean albumin extraction methodologies used in this study resulted in mung bean albumin extracts with different protein profiles (Figures 2 and 3). Only the albumin isolate obtained from mung bean albumins with borate buffer pH 8.3 had no 40–50 kDa globulins (Figure 2, lane 6). Therefore, the albumin isolate obtained from the borate buffer pH 8.3 extraction methodology was determined to be purer than the other protein isolates analyzed. Since mung bean albumins are not currently known to be larger than 30 kDa, the 30 kDa permeate of the albumin isolate obtained from the borate buffer pH 8.3 extraction protocol was sequenced. In addition, it was used for all antioxidant assays performed, as this isolate was determined to be the most consistent with the current literature [19,20]. The protein profile of this permeate is given on lane 9 of Figure 2.

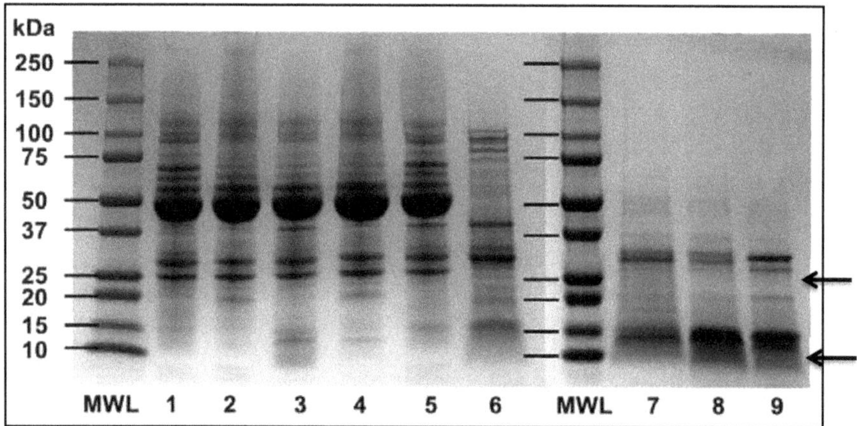

Figure 2. SDS-PAGE (Sodium Dodecyl Sulfate–Polyacrylamide Gel Electrophoresis) of mung bean albumin extracts and 30 kDa MWCO (molecular weight cut-off) filtrates. MWL = molecular weight ladder. (**1**) Water pH = 7.5 extract, (**2**) Albumin isolate from water pH = 7.5 extract, (**3**) Water pH = 3.0 extract, (**4**) Albumin isolate from water pH = 3.0 extract, (**5**) Borate buffer pH = 8.3 extract, (**6**) Albumin isolate from borate buffer pH = 8.3 extract. Samples on lanes (**2**), (**4**) and (**6**) were filtered using a 30 kDa MWCO centrifugal filter to produce the permeates on (**7**), (**8**) and (**9**), respectively. Arrows indicate the protein fractions that are most likely to be mung bean albumin according to their molecular weight.

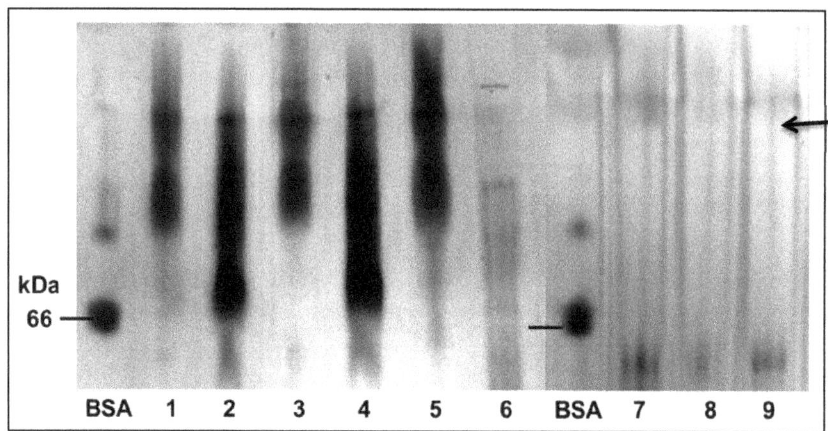

Figure 3. The native Blue-PAGE of mung bean albumin extracts and 30 kDa MWCO filtrates. BSA = bovine serum albumin. BSA was used as a reference protein. (**1**) Water pH = 7.5 extract, (**2**) Albumin isolate from water pH = 7.5 extract, (**3**) Water pH = 3.0 extract, (**4**) Albumin isolate from water pH = 3.0 extract, (**5**) Borate buffer pH = 8.3 extract, (**6**) Albumin isolate from borate buffer pH = 8.3 extract. Samples on lanes (**2**), (**4**) and (**6**) were filtered using a 30 kDa MWCO centrifugal filter to produce the permeates on (**7**), (**8**) and (**9**), respectively. Arrows indicate the protein fractions that are most likely to be mung bean albumin according to their molecular weight.

3.2. Potentially Antioxidant Peptides from Mung Bean Albumins by In Silico Hydrolysis

For the 10 kDa mung bean albumin sequence Q9FRT8, the bioactive fragments LK and FC were found for the sequence using the BIOPEP Database. These fragments were found in the Q9FRT8 sequence three times, and their locations are shown in Figure 4. For the 30 kDa mung bean albumin sequence Q43680, the bioactive fragments HL, AY, EL, IR, LK, KP, VY and FC were found using the

BIOPEP Database. These fragments were found in the Q43680 sequence 11 times, and their locations are shown in Figure 5.

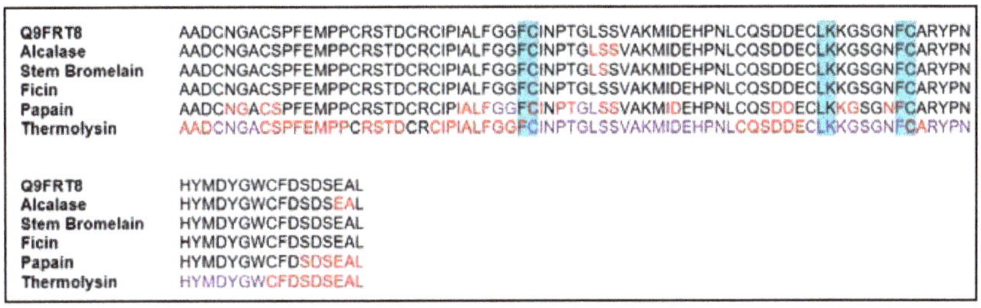

```
Q9FRT8          AADCNGACSPFEMPPCRSTDCRCIPIALFGGFCINPTGLSSVAKMIDEHPNLCQSDDECLKKGSGNFCARYPN
Alcalase        AADCNGACSPFEMPPCRSTDCRCIPIALFGGFCINPTGLSSVAKMIDEHPNLCQSDDECLKKGSGNFCARYPN
Stem Bromelain  AADCNGACSPFEMPPCRSTDCRCIPIALFGGFCINPTGLSSVAKMIDEHPNLCQSDDECLKKGSGNFCARYPN
Ficin           AADCNGACSPFEMPPCRSTDCRCIPIALFGGFCINPTGLSSVAKMIDEHPNLCQSDDECLKKGSGNFCARYPN
Papain          AADCNGACSPFEMPPCRSTDCRCIPIALFGGFCINPTGLSSVAKMIDEHPNLCQSDDECLKKGSGNFCARYPN
Thermolysin     AADCNGACSPFEMPPCRSTDCRCIPIALFGGFCINPTGLSSVAKMIDEHPNLCQSDDECLKKGSGNFCARYPN

Q9FRT8          HYMDYGWCFDSDSEAL
Alcalase        HYMDYGWCFDSDSEAL
Stem Bromelain  HYMDYGWCFDSDSEAL
Ficin           HYMDYGWCFDSDSEAL
Papain          HYMDYGWCFDSDSEAL
Thermolysin     HYMDYGWCFDSDSEAL
```

Figure 4. Locations of possible hydrolysis, by proteases selected, within the mung bean albumin sequence Q9FRT8 of the UniProt Database. Amino acids in the sequence are coded using their one letter abbreviation. Hydrolysis is expected to occur between any two adjacent amino acids color-coded the same color (red or purple), but not between amino acids color-coded with different colors or in black text. Each block represents a part of the sequence depicting the different specificities of the proteases stated in the left column. The top block is the start of the sequence. Antioxidant fragments according to the BIOPEP Database in the sequence are highlighted in blue.

```
Q43680          MSNLPYINAAFRFSSRDYEVYFFAKNKYVRLQYTPGKTEDKILTNLRLISSGFPSLAGTPFAEPGIDSAFHTE
Alcalase        MSNLPYINAAFRFSSRDYEVYFFAKNKYVRLQYTPGKTEDKILTNLRLISSGFPSLAGTPFAEPGIDSAFHTE
Stem Bromelain  MSNLPYINAAFRFSSRDYEVYFFAKNKYVRLQYTPGKTEDKILTNLRLISSGFPSLAGTPFAEPGIDSAFHTE
Ficin           MSNLPYINAAFRFSSRDYEVYFFAKNKYVRLQYTPGKTEDKILTNLRLISSGFPSLAGTPFAEPGIDSAFHTE
Papain          MSNLPYINAAFRFSSRDYEVYFFAKNKYVRLQYTPGKTEDKILTNLRLISSGFPSLAGTPFAEPGIDSAFHTE
Thermolysin     MSNLPYINAAFRFSSRDYEVYFFAKNKYVRLQYTPGKTEDKILTNLRLISSGFPSLAGTPFAEPGIDSAFHTE

Q43680          ASEAYVFSANNRAYIDYAPGTTNDKILAGPTTIAEMFPVLRNTVFADSIDSAFRSTKGKEVYLFKGNKYVRID
Alcalase        ASEAYVFSANNRAYIDYAPGTTNDKILAGPTTIAEMFPVLRNTVFADSIDSAFRSTKGKEVYLFKGNKYVRID
Stem Bromelain  ASEAYVFSANNRAYIDYAPGTTNDKILAGPTTIAEMFPVLRNTVFADSIDSAFRSTKGKEVYLFKGNKYVRID
Ficin           ASEAYVFSANNRAYIDYAPGTTNDKILAGPTTIAEMFPVLRNTVFADSIDSAFRSTKGKEVYLFKGNKYVRID
Papain          ASEAYVFSANNRAYIDYAPGTTNDKILAGPTTIAEMFPVLRNTVFADSIDSAFRSTKGKEVYLFKGNKYVRID
Thermolysin     ASEAYVFSANNRAYIDYAPGTTNDKILAGPTTIAEMFPVLRNTVFADSIDSAFRSTKGKEVYLFKGNKYVRID

Q43680          YDSKQLVGSIRNISDGFPVLNGTGFESGIDASFASHKEPEAYLFKGDKYVRIHFTPGKTDDTLVGDVRPILD
Alcalase        YDSKQLVGSIRNISDGFPVLNGTGFESGIDASFASHKEPEAYLFKGDKYVRIHFTPGKTDDTLVGDVRPILD
Stem Bromelain  YDSKQLVGSIRNISDGFPVLNGTGFESGIDASFASHKEPEAYLFKGDKYVRIHFTPGKTDDTLVGDVRPILD
Ficin           YDSKQLVGSIRNISDGFPVLNGTGFESGIDASFASHKEPEAYLFKGDKYVRIHFTPGKTDDTLVGDVRPILD
Papain          YDSKQLVGSIRNISDGFPVLNGTGFESGIDASFASHKEPEAYLFKGDKYVRIHFTPGKTDDTLVGDVRPILD
Thermolysin     YDSKQLVGSIRNISDGFPVLNGTGFESGIDASFASHKEPEAYLFKGDKYVRIHFTPGKTDDTLVGDVRPILD

Q43680          GWPVLKAFCLCELNKPSLSCINHLSLVTINKAFISNVCLFFFNVTLGLEACFLS
Alcalase        GWPVLKAFCLCELNKPSLSCINHLSLVTINKAFISNVCLFFFNVTLGLEACFLS
Stem Bromelain  GWPVLKAFCLCELNKPSLSCINHLSLVTINKAFISNVCLFFFNVTLGLEACFLS
Ficin           GWPVLKAFCLCELNKPSLSCINHLSLVTINKAFISNVCLFFFNVTLGLEACFLS
Papain          GWPVLKAFCLCELNKPSLSCINHLSLVTINKAFISNVCLFFFNVTLGLEACFLS
Thermolysin     GWPVLKAFCLCELNKPSLSCINHLSLVTINKAFISNVCLFFFNVTLGLEACFLS
```

Figure 5. Locations of possible hydrolysis, by proteases selected, within the mung bean albumin sequence Q43680 of the UniProt Database. Amino acids in the sequence are coded using their one letter abbreviation. Hydrolysis is expected to occur between any two adjacent amino acids color-coded the same color (red or purple), but not between amino acids color-coded with different colors or in black text. Each block represents a part of the sequence depicting the different specificities of the proteases stated in the left column. The top block is the start of the sequence. Antioxidant fragments according to the BIOPEP Database in the sequence are highlighted in blue.

Thermolysin was found to be the protease most likely to produce small bioactive di-/tri-peptides from both mung bean albumin sequences in the UniProt Database analyzed (Table 2). Thermolysin was also found to be the enzyme most likely to destroy bioactive fragments in the mung bean albumin sequences analyzed, but given that it could hydrolyze the majority of the mung bean albumin sequences, this was expected (Figures 4 and 5). Thermolysin was selected to hydrolyze mung bean albumins in vitro due to its greater potential to produce antioxidant peptides.

Table 2. Possible Small Antioxidant Peptides Produced from the Theoretical Hydrolysis of Mung Bean Albumin Sequences by Papain and Thermolysin.

Protease	Protein Sequence	Potentially Antioxidant Di-/Tri-Peptides Expected to be Produced [1]	Number of Potentially Broken Antioxidant Fragments
Papain	Q9FRT8	GFC	0
	Q43680	EVY, VY, VYF, EAY, AY, AYV, VLK, KPS	6
Thermolysin	Q9FRT8	GFC,FC, FCI,LK, LKK, NFC	1
	Q43680	EVY, VY, VYF, EAY, AY, AYV, AYI, VYL, SIR,IR,IRN, AYL, VLK,LK, LKA, AFC, FC, EL, ELN, NKP, KP, KPS, NHL,HL, HLS	9

[1] Potentially antioxidant fragments in the analyzed di-/tri-peptides are labeled in blue.

3.3. Antioxidant Activity of Mung Bean Albumin Hydrolysates and Peptide Sequencing

Results in Table 3 and Figure S1 show that mung bean albumins had some antioxidant capacity themselves (t = 0). Thermolysin and pepsin–pancreatin hydrolysis increased the antioxidant capacity in terms of ferrous ion chelating activity with significant changes due to time of hydrolysis. The ferrous ion chelating activity of the hydrolysates derived via both thermolysin and pepsin–pancreatin enzymatic hydrolysis was generally higher than the ABTS radical scavenging activity. ABTS and ORAC values did not present a significant increase in the antioxidant activities of any of the hydrolysates produced by the enzymes tested. The ferrous ion chelating activities of thermolysin and gastrointestinal peptides hydrolysates were not statistically different at the end of the hydrolysis. The same pattern was observed for ORAC values between thermolysin and gastrointestinal peptides. There was, however, significant difference between t = 0 and the rest of the time points for the ferrous ion chelating activity of thermolysin-derived peptides, except for pepsin, where the activity decreased. There was no significant difference in the ABTS radical scavenging activity at different time points for both thermolysin-derived and simulated gastrointestinal digestion-derived peptides.

Table 3. Antioxidant Potential of Mung Bean Hydrolysates with Varying Hydrolysis Times.

Enzyme Scheme	Time (min)	ABTS Radical Scavenging Activity (Ascorbic Acid Equivalent, µM)	Iron Chelating Activity (EDTA Equivalent, µM)	ORAC (Trolox Equivalent, µM)
Thermolysin + MBA	0	685.0 ± 333.8	3034.8 ± 216.4	127 ± 19
	5	632.7± 288.6	5334.9 ± 45.3	125 ± 21
	10	683.2 ± 296.1	5337.8 ± 38.3	115 ± 17
	25	449.7 ± 301.4	5328.1 ± 17.4	98 ± 39
	45	511.0 ± 255.7	5329.2 ± 35.4	116 ± 17
	60	493.9 ± 297.7	5336.1 ± 26.2	127 ± 17
	240	646.2 ± 324.7	5161.7 ± 280.2	127 ± 17
Pepsin + MBA	0	513.7 ± 230.4	3005.5 ± 248.0	122 ± 17
	15	533.5 ± 332.0	472.4 ± 227.4	126 ± 8
	25	535.3 ± 328.0	529.4 ± 234.1	96 ± 20
	45	540.8 ± 307.3	500.2 ± 173.3	99 ± 35
	60	477.7 ± 316.6	490.7 ± 300.4	122 ± 12
	90	638.1 ± 340.2	594.3 ± 108.6	107 ± 42
	120	495.7 ± 342.3	434.3 ± 230.1	101 ± 46
Pepsin-Pancreatin + MBA	0	487.6 ± 338.4	448.5 ± 240.1	145 ± 38
	15	450.2 ± 283.0	4924.6 ± 119.4	105 ± 5
	25	474.1 ± 245.5	4901.0 ± 95.0	97 ± 31
	45	559.7 ± 197.0	4886.9 ± 182.8	121 ± 17
	60	456.9 ± 260.3	4884.4 ± 105.6	125 ± 16
	90	552.9 ± 256.4	4794.9 ± 16.5	124 ± 16
	120	581.3 ± 242.9	4843.5 ± 155.5	127 ± 15

MBA = Mung Bean Albumin; 1.0 mg/mL of EDTA is equivalent to 3421.8 µM; 1.5 mg/mL of EDTA is equivalent to 5132.8 µM; 0.15 mg/mL of ascorbic acid is equivalent to 851.7 µM; 1.0 mg/mL of ascorbic acid is equivalent to 5677.9 µM.

Results in Tables 4 and 5 also show that the peptides derived from enzymatic hydrolysis using gastrointestinal enzymes (pepsin-pancreatin), and with antioxidant potential, have higher molecular mass (~328 Da) compared to peptides derived from enzymatic hydrolysis using thermolysin (~253 Da). Enzymatic hydrolysis using thermolysin also produced more peptides with high hydrophobicity (>10 kcal/mol). All the peptides reported in Tables 4 and 5 have been found to have antioxidant potential according to the BIOPEP Database, which is compounded from previously published literature.

Table 4. Sequences and Functional Properties of Mung Bean Albumin Peptides Derived from Simulated Gastrointestinal Digestion.

Peptide Sequence	Molecular Mass (Da)	Hydrophobicity (Kcal/mol)	Isoelectric Point	Charge
MD	264	10.87	2.95	−1
QSA	304	9.63	5.49	0
EW	333	9.44	3.27	−1
LGW	374	5.71	5.69	0
KK	274	13.5	10.57	2
SVP	301	8.04	5.18	0
DVAF	450	9.87	3.05	−1

Table 5. Sequences and Properties of Mung Bean Albumin Peptides Derived from Thermolysin Enzymatic Hydrolysis.

Peptide Sequence	Molecular Mass (Da)	Hydrophobicity (Kcal/mol)	Isoelectric Point	Charge
KK	274	14.50	10.57	2
DM	264	10.87	3.02	−1
SY	268	7.65	5.38	0
W	204	5.81	5.64	0

Supplementary Materials Table S1 presents the complete sequences and functional properties of mung bean albumin hydrolysates derived from gastrointestinal enzymatic hydrolysis, and the effect of thermolysin. The values of the peptides found ranged between 236 and 1509 kDa, 5.81 and 20.74 Kcal/mol and 2.82 and 11.18 for molecular mass, hydrophobicity and isoelectric point, respectively. Supplementary Materials Table S2 presents the peptide sequences, functional properties and bioactivities of mung bean albumin hydrolysates derived from thermolysin enzymatic hydrolysis.

4. Discussion

A defining characteristic of mung bean albumins is their high solubility in water, but extracting them with water alone was found to be inadequate to isolate them from other mung bean proteins, as demonstrated by the SDS-PAGE result in Figure 2. The extraction of mung bean albumins using sodium borate buffer (pH = 8.3) and then subsequent centrifugal filtration was more effective for obtaining isolated mung bean albumins that were within the currently known filtration 3–30 kDa mass range. At pH 7.0, other mung bean proteins aside from albumins were soluble [28], and therefore the extraction of mung bean albumin with water alone at pH 7.0 will inevitably induce the extraction of other proteins. The solubility of other mung bean proteins lowered as the pH became higher than 7.0, while the solubility of albumin increased as the pH neared 8.0. Thus, using a buffer of pH 8.3 to extract mung bean albumins was more effective than using pure water, as the difference in solubility between albumin and the rest of the proteins at that pH allowed mung bean albumins to be effectively isolated. Mung bean albumin peptides produced through hydrolysis by thermolysin and gastrointestinal enzymes showed antioxidant potential. Mung bean albumin peptides showed high ferrous ion chelating activity, but low ABTS radical scavenging activity (Supplementary Materials Figure S1). The ferrous ion chelating activity of mung bean albumin peptides derived from pepsin–pancreatin enzymatic hydrolysis was found to be higher than the ferrous ion chelating activity of cowpea and common bean protein hydrolysates derived from the same enzyme scheme, as reported by Segura-Campos et al. [36]. The ferrous ion chelating activity of mung bean albumin peptides was also found to be higher than the ferrous ion chelating activity of soybean lunasin, as reported by Garcia-Nebot et al. [37]. It was also found to have higher ferrous ion chelating activity compared to the Bambara groundnut (*Vigna subterranea*) protein hydrolysates, as reported by Arise et al. [38]. It also has a much higher ferrous ion chelating activity compared to phaseolin and bean protein hydrolysates, as reported by Carrasco-Castilla et al. [39] However, it has lower ferrous ion chelating activity compared to pea (*Pisum sativum* L.) protein hydrolysates, as reported by Pownall et al. [40] Thus, it can be said that mung bean albumin peptides are effective ferrous ion chelators. Of note, food-safe commercial enzymes such as alcalase and flavourzyme have been reported to produce hydrolysates with higher ferrous ion chelating activity compared to pepsin–pancreatin, as demonstrated by the sources above. These results are similar in this study, wherein thermolysin was found to produce peptides with higher iron ion chelating activity compared to pepsin–pancreatin, although the difference was also found to be not statistically significant.

Ascorbic acid was used as a standard in the ABTS assay as it is a water-soluble antioxidant present in foods. Previous publications also report the use of ascorbic acid as a standard for the ABTS assay [41].

The iron chelating and radical scavenging activities showed variation at different hydrolysis times for peptides derived via thermolysin and those derived via pepsin–pancreatin enzymatic hydrolysis (Figure S1). Overall, mung bean albumin peptides produced through hydrolysis by thermolysin showed higher radical scavenging and metal chelating activity, although the difference was not statistically significant at the end of hydrolysis. However, both types of peptides showed similar ORAC values.

According to the study on mung bean protein hydrolysates done by Sonklin et al. [42], radical scavenging activity was affected by both enzyme concentration and the hydrolysis time of a hydrolysate. Increasing enzyme concentration and hydrolysis time increased radical scavenging activity, but only until reaching a critical point, after which the activity became constant and decreased. As we used the same concentration of enzyme throughout the hydrolysis process, the variation in antioxidant capacity can be explained by the difference in hydrolysis time. Budseekoad et al. [26] reported that the iron ion binding capabilities of mung bean protein hydrolysates vary with hydrolysis time, and that different enzymes produce peptides with the highest calcium ion binding capabilities at different times. It can be theorized that different peptides with different antioxidant capacities were produced at different hydrolysis times, as was observed by this investigation.

Studies performed by Kong and Xiong [43] and Ajibola et al. [44] revealed that hydrophobicity played a part in the antioxidant capacity of peptides, with more hydrophobic peptides exhibiting a higher antioxidant capacity. Enzymatic hydrolysis using thermolysin produced mostly peptides with high hydrophobicity (>10 kcal/mol) compared to enzymatic hydrolysis using gastrointestinal enzymes, which explains the difference in the antioxidant capacity between the two types of hydrolysates analyzed. According to Zhu et al. [45], the size of peptides is a significant factor in their antioxidant capacity, whereby peptides with lower molecular mass exhibit higher antioxidant capacities. This assessment agreed with our results, as the thermolysin enzymatic hydrolysis produced peptides with lower molecular mass (Table 5).

Both high hydrophobicity and small molecular weight are factors that increase the absorption of peptides in the small intestine [46]. Peptides produced with thermolysin were of small molecular weight and of high hydrophobicity, indicating that they are expected to be absorbed during digestion. Antioxidant and hydrophobic casein peptides have already been shown to have good bioavailability using a Caco-2 cell model [47]. Future cellular studies are needed to measure the antioxidant capacity of mung bean protein-derived peptides before animal and human studies are performed.

5. Conclusions

Mung bean albumins at pH 7.0 have the same solubility as other mung bean proteins, making their extraction using pure water ineffective in isolating mung bean albumins from other proteins. Mung bean albumins at pH 8.3 had higher solubility than other mung bean proteins, making this pH better for mung bean albumin extraction. Mung bean albumin hydrolysates showed antioxidant potential in terms of ferrous ion chelating and ORAC values. They are particularly effective ferrous ion chelators. Hydrolysates produced via thermolysin enzymatic hydrolysis had higher antioxidant capacity overall due to their high hydrophobicity and low molecular mass. High hydrophobicity and low molecular mass are two factors that can increase the intestinal absorption of thermolysin-derived mung bean albumin peptides, but more in vivo studies are required to quantify their bioavailability in humans. The variation of antioxidant capacity over different time points during hydrolysis showed that different hydrolysis times produced different peptides of different antioxidant capacities. To our knowledge, this is the first study to investigate the antioxidant potential of mung bean albumin peptides through a variety of methods.

Supplementary Materials: The following are available online at http://www.mdpi.com/2304-8158/9/9/1241/s1, Table S1: Sequences, Functional Properties and Bioactivity of Mung Bean Albumin Hydrolysates Derived from Gastrointestinal Enzymatic Hydrolysis; Table S2: Sequences, Functional Properties and Bioactivity of Mung Bean Albumin Hydrolysates Derived from Thermolysin Enzymatic Hydrolysis; Figure S1: Antioxidant Potential of Mung Bean Albumin Hydrolysates at Various Hydrolysis Times from Two Different Enzyme Schemes. The star marks the time point at which pancreatin was added 2 h after pepsin. These results were obtained from hydrolysates derived from an average of 576 µg/mL of mung bean albumin.

Author Contributions: Conceptualization, J.K., E.G.d.M.; Methodology, L.M.R.H., J.K.; Validation L.M.R.H., J.K.; Formal Analysis, J.K., L.M.R.H.; Investigation, J.K.; Resources, E.G.d.M.; Data Curation, J.K., L.M.R.H.; Writing—original draft preparation, J.K., L.M.R.H.; Writing—review and editing, J.K., L.M.R.H., E.G.d.M.; Visualization, J.K., L.M.R.H., E.G.d.M.; Supervision, E.G.d.M.; Project Administration, E.G.d.M.; Funding acquisition, J.K., E.G.d.M. All authors have read and agreed to the published version of the manuscript.

Funding: This research was funded by USDA-NIFA 1014457 to EDM and the Undergraduate Research program of College of Agricultural, Consumer and Environmental Sciences to JK, University of Illinois at Urbana-Champaign.

Acknowledgments: The authors thank Diego Luna-Vital for his help during the course of this study and Erick Damian Castañeda Reyes for his help in the data analysis section.

Conflicts of Interest: The authors declare no conflict of interest.

References

1. Oo, Z.Z.; Ko, L.T.; Than, S.S. Physico-chemical properties of extracted mung bean protein concentrate. *Am. J. Food Sci. Technol.* **2017**, *5*, 265–269.

2. Gunarti, D.R.; Rahmi, H.; Sadikin, M. Isolation and purification of thiamine binding protein from mung bean. *HAYATI J. Biosci.* **2013**, *20*, 1–6. [CrossRef]

3. Tang, D.; Dong, Y.; Ren, H.; Li, L.; He, C. A review of phytochemistry, metabolite changes, and medicinal uses of the common food mung bean and its sprouts (*Vigna radiata*). *Chem. Cent. J.* **2014**, *8*, 4. [CrossRef] [PubMed]

4. Mendoza, E.M.; Adachi, M.; Bernardo, A.E.; Utsumi, S. Mungbean [*Vigna radiata* (L.) Wilczek] globulins: Purification and characterization. *J. Agric. Food Chem.* **2001**, *49*, 1552–1558. [CrossRef] [PubMed]

5. Wang, M.; Jiang, L.; Li, Y.; Liu, Q.; Wang, S.; Sui, X. Optimization of extraction process of protein isolate from mung bean. *Procedia Eng.* **2011**, *15*, 5250–5258. [CrossRef]

6. Brishti, F.H.; Zarei, M.; Muhammad, S.K.S.; Ismail-Fitry, M.R.; Shukri, R.; Saari, N. Evaluation of the functional properties of mung bean protein isolate for development of textured vegetable protein. *Int. Food Res. J.* **2017**, *24*, 1595–1605.

7. Shi, Z.; Yao, Y.; Zhu, Y.; Ren, G. Nutritional composition and antioxidant activity of twenty mung bean cultivars in China. *Crop J.* **2016**, *4*, 398–406. [CrossRef]

8. Deraz, S.F.; Khalil, A.A. Strategies to improve protein quality and reduce antinutritional factors in mung bean. *Food* **2008**, *2*, 25–38.

9. Anwar, F.; Iqbal, Z.; Sultana, B.; Iqbal, S.; Saari, N. Effects of extraction system on antioxidant attributes of mungbean [*Vigna radiata* (L.) Wilczek]. *Int. J. Food Prop.* **2013**, *16*, 527–535. [CrossRef]

10. Varma, A.; Mishra, S.P.; Tripathi, A.; Shukla, U.K. Biochemical composition and storage protein profiling of mungbean (*Vigna radiata* L. wilczek) cultivars. *J. Pharmacog. Phytochem.* **2018**, *7*, 708–713.

11. Gupta, N.; Srivastava, N.; Bhagyawant, S.S. Vicilin—A major storage protein of mungbean exhibits antioxidative potential, antiproliferative effects and ACE inhibitory activity. *PLoS ONE* **2018**, *13*, e0191265. [CrossRef] [PubMed]

12. Wang, J.; Tse, Y.C.; Hinz, G.; Robinson, D.G.; Jiang, L. Storage globulins pass through the Golgi apparatus and multivesicular bodies in the absence of dense vesicle formation during early stages of cotyledon development in mung bean. *J. Exp. Bot.* **2012**, *63*, 1367–1380. [CrossRef] [PubMed]

13. Yi-Shen, Z.; Shuai, S.; Fitzgerald, R. Mung bean proteins and peptides: Nutritional, functional and bioactive properties. *Food Nutr. Res.* **2018**, *62*, 1–11. [CrossRef] [PubMed]

14. Li, S.; Cao, Y.; Geng, F. Genome-wide identification and comparative analysis of albumin family in vertebrates. *Evol. Bioinform. Online* **2017**, *13*, 1176934317716089. [CrossRef]

15. Moreno, F.J.; Clemente, A. 2S Albumin storage proteins: What makes them food allergens? *Open Biochem. J.* **2008**, *2*, 16–28. [CrossRef] [PubMed]

16. Galbas, M.; Porzucek, F.; Woźniak, A.; Slomski, R.; Selwet, M. Isolation of low-molecular albumins of 2S fraction from soybean (*Glycine max* (L.) Merrill). *Acta Biochim. Pol.* **2013**, *60*, 107–110. [CrossRef] [PubMed]

17. Scarafoni, A.; Gualtieri, E.; Barbiroli, A.; Carpen, A.; Negri, A.; Duranti, M. Biochemical and functional characterization of an albumin protein belonging to the hemopexin superfamily from *Lens culinaris* seeds. *J. Agric. Food Chem.* **2011**, *59*, 9637–9644. [CrossRef]

18. Kortt, A.A. Isolation and characterisation of a major seed albumin. *Int. J. Peptide Prot. Res.* **1986**, *28*, 613–619. [CrossRef]

19. Kang, Y.J.; Kim, S.K.; Kim, M.Y.; Lestari, P.; Kim, K.H.; Ha, B.K.; Jun, T.H.; Hwang, W.J.; Lee, T.; Lee, J.; et al. Genome sequence of mungbean and insights into evolution within Vigna species. *Nat. Commun.* **2014**, *5*, 5443. [CrossRef]

20. Yamazaki, T.; Takaoka, M.; Katoh, E.; Hanada, K.; Sakita, M.; Sakata, K.; Nishiuchi, Y.; Hirano, H. A possible physiological function and the tertiary structure of a 4-kDa peptide in legumes. *Eur. J. Biochem.* **2003**, *270*, 1269–1276. [CrossRef]

21. Kou, X.; Gao, J.; Zhang, Z.; Wang, H.; Wang, X. Purification and identification of antioxidant peptides from chickpea (Cicer arietinum L.) albumin hydrolysates. *LWT Food Sci. Technol.* **2013**, *50*, 591–598. [CrossRef]

22. Xue, Z.; Wen, H.; Zhai, L.; Yu, Y.; Li, Y.; Yu, W.; Cheng, A.; Wang, C.; Kou, X. Antioxidant activity and anti-proliferative effect of a bioactive peptide from chickpea (*Cicer arietinum* L.). *Food Res. Int.* **2015**, *77*, 75–81. [CrossRef]

23. Benedé, S.; Molina, E. Chicken egg proteins and derived peptides with antioxidant properties. *Foods* **2020**, *9*, 735. [CrossRef] [PubMed]

24. Lule, V.K.; Garg, S.; Pophaly, S.D.; Tomar, S.K. Potential health benefits of lunasin: A multifaceted soy-derived bioactive peptide. *J. Food Sci.* **2015**, *80*, R485–R494. [CrossRef]

25. Hernández-Ledesma, B.; Hsieh, C.; De Lumen, B.O. Biochemical and biophysical research communications antioxidant and anti-inflammatory properties of cancer preventive peptide lunasin in RAW 264.7 macrophages. *Biochem. Biophys. Res. Commun.* **2009**, *390*, 803–808. [CrossRef]

26. Budseekoad, S.; Yupanqui, C.T.; Sirinupong, N.; Alashi, A.M.; Aluko, R.E.; Youravong, W. Structural and functional characterization of calcium and iron-binding peptides from mung bean protein hydrolysate. *J. Funct. Foods* **2018**, *49*, 333–341. [CrossRef]

27. Lapsongphon, N.; Yongsawatdigul, J. Production and purification of antioxidant peptides from a mungbean meal hydrolysate by *Virgibacillus* sp. SK37 proteinase. *Food Chem.* **2013**, *141*, 992–999. [CrossRef]

28. Du, M.; Xie, J.; Gong, B.; Xu, X.; Tang, W.; Li, X.; Li, C.; Xie, M. Extraction, physicochemical characteristics and functional properties of Mung bean protein. *Food Hydrocoll.* **2018**, *76*, 131–140. [CrossRef]

29. Singh, D.K.; Rao, A.S.; Singh, R.; Jambunathan, R. Amino acid composition of storage proteins of a promising chickpea (*Cicer arietinum* L) cultivar. *J. Sci. Food Agric.* **1988**, *43*, 373–379. [CrossRef]

30. De Souza Rocha, T.; Hernandez, L.M.R.; Chang, Y.K.; de Mejía, E.G. Impact of germination and enzymatic hydrolysis of cowpea bean (Vigna unguiculata) on the generation of peptides capable of inhibiting dipeptidyl peptidase IV. *Food Res. Int.* **2014**, *64*, 799–809. [CrossRef]

31. Daniel, H. Molecular and integrative physiology of intestinal peptide transport. *Annu. Rev. Physiol.* **2004**, *66*, 361–384. [CrossRef] [PubMed]

32. Mojica, L.; de Mejía, E.G. Optimization of enzymatic production of anti-diabetic peptides from black bean (*Phaseolus vulgaris* L.) proteins, their characterization and biological potential. *Food Funct.* **2016**, *7*, 713–727. [CrossRef] [PubMed]

33. Held, P.; Brescia, P.J. Determining Antioxidant Potential Using an Oxygen Radical Absorbance Capacity (ORAC) Assay with SynergyTM H4. Available online: https://www.labmanager.com/products-in-action/determining-antioxidant-potential-using-an-oxygen-radical-absorbance-capacity-orac-assay-10182 (accessed on 1 June 2020).

34. Grancieri, M.; Martino, H.S.D.; Gonzalez de Mejia, E. Digested total protein and protein fractions from chia seed (Salvia hispanica L.) had high scavenging capacity and inhibited 5-LOX, COX-1-2, and iNOS enzymes. *Food Chem.* **2019**, *289*, 204–214. [CrossRef] [PubMed]

35. PepDraw. Available online: https://www.tulane.edu/~{}biochem/WW/PepDraw/index.html (accessed on 30 July 2019).

36. Segura-Campos, M.; Ruiz-Ruiz, J.; Chel-guerrero, L.; Betancur-Ancona, D. Antioxidant activity of *Vigna unguiculata* L. walp and hard-to-cook *Phaseolus vulgaris* L. protein hydrolysates. *CyTA J. Food* **2013**, *11*, 208–215. [CrossRef]

37. García-Nebot, M.J.; Recio, I.; Hernández-ledesma, B. Antioxidant activity and protective effects of peptide lunasin against oxidative stress in intestinal Caco-2 cells. *Food Chem. Tox.* **2014**, *65*, 155–161. [CrossRef]

38. Arise, A.K.; Alashi, A.M.; Nwachukwu, I.D. Function (*Vigna subterranea*) protein hydrolysates and. *Food Funct.* **2016**, *7*, 2431–2437. [CrossRef]

39. Carrasco-Castilla, J.; Hernández-Alvarez, A.J.; Jiménez-Martínez, C.; Jacinto-Hernández, C.; Alaiz, M.; Girón-Calle, J.; Vioque, J.; Dávila-Ortiz, G. Antioxidant and metal chelating activities of peptide fractions from phaseolin and bean protein hydrolysates. *Food Chem.* **2012**, *135*, 1789–1795. [CrossRef]

40. Pownall, T.L.; Udenigwe, C.C.; Aluko, R.E. Amino acid composition and antioxidant properties of pea seed (*Pisum sativum* L.) enzymatic protein hydrolysate fractions. *J. Agric. Food Chem.* **2010**, *58*, 4712–4718. [CrossRef]

41. Tang, X.; He, Z.; Dai, Y.; Xiong, Y.L.; Xie, M.; Chen, J. Peptide fractionation and free radical scavenging activity of zein hydrolysate. *J. Agric. Food Chem.* **2010**, *58*, 587–593. [CrossRef]

42. Sonklin, C.; Laohakunjit, N.; Kerdchoechuen, O. Assessment of antioxidant properties of membrane ultrafiltration peptides from mungbean meal protein hydrolysates. *PeerJ* **2018**, *6*, e5337. [CrossRef]

43. Kong, B.; Xiong, Y.L. Antioxidant activity of zein hydrolysates in a liposome system and the possible mode of action. *J. Agric. Food Chem.* **2006**, *54*, 6059–6068. [CrossRef] [PubMed]

44. Ajibola, C.F.; Fashakin, J.B.; Fagbemi, T.N.; Aluko, R.E. Effect of Peptide Size on Antioxidant Properties of African Yam Bean Seed (*Sphenostylis stenocarpa*) Protein Hydrolysate Fractions. *Int. J. Mol. Sci.* **2011**, *12*, 6685–6702. [CrossRef] [PubMed]

45. Zhou, L.; Chen, J.; Tang, X.; Xiong, Y.L. Reducing, radical scavenging, and chelation properties of in vitro digests of alcalase-treated zein hydrolysate. *J. Agric. Food Chem.* **2008**, *56*, 2714–2721. [CrossRef] [PubMed]

46. Xu, Q.; Hong, H.; Wu, J.; Yan, X. Bioavailability of bioactive peptides derived from food proteins across the intestinal epithelial membrane: A review. *Trends Food Sci. Technol.* **2019**, *86*, 399–411. [CrossRef]

47. Wang, B.; Li, B. Charge and hydrophobicity of casein peptides influence transepithelial transport and bioavailability. *Food Chem.* **2018**, *245*, 646–652. [CrossRef] [PubMed]

Article

In Silico and In Vitro Analysis of Multifunctionality of Animal Food-Derived Peptides

Lourdes Amigo [1], Daniel Martínez-Maqueda [2] and Blanca Hernández-Ledesma [1,*]

[1] Departamento de Bioactividad y Análisis de Alimentos, Instituto de Investigación en Ciencias de la Alimentación (CIAL, CSIC-UAM, CEI-UAM+CSIC), Nicolás Cabrera 9, 28049 Madrid, Spain; lourdes.amigo@csic.es

[2] Departamento de Investigación Agroalimentaria, Instituto Madrileño de Investigación y Desarrollo Rural, Agrario y Alimentario (IMIDRA), 28800 Madrid, Spain; daniel.martinez.maqueda@madrid.org

* Correspondence: b.hernandez@csic.es; Tel.: +34-001-70-970

Received: 6 July 2020; Accepted: 21 July 2020; Published: 24 July 2020

Abstract: Currently, the associations between oxidative stress, inflammation, hypertension, and metabolic disturbances and non-communicable diseases are very well known. Since these risk factors show a preventable character, the searching of food peptides acting against them has become a promising strategy for the design and development of new multifunctional foods or nutraceuticals. In the present study, an integrated approach combining an in silico study and in vitro assays was used to confirm the multifunctionality of milk and meat protein-derived peptides that were similar to or shared amino acids with previously described opioid peptides. By the in silico analysis, 15 of the 27 assayed peptides were found to exert two or more activities, with Angiotensin-converting enzyme (ACE) inhibitory, antioxidant, and opioid being the most commonly found. The in vitro study confirmed ACE-inhibitory and antioxidant activities in 15 and 26 of the 27 synthetic peptides, respectively. Four fragments, RYLGYLE, YLGYLE, YFYPEL, and YPWT, also demonstrated the ability to protect Caco-2 and macrophages RAW264.7 cells from the oxidative damage caused by chemicals. The multifunctionality of these peptides makes them promising agents against oxidative stress-associated diseases.

Keywords: bioactive peptides; animal protein; multifunctionality; antioxidant activity; in silico; cell models

---

1. Introduction

Non-communicable diseases (NCDs) such as cardiovascular and neurodegenerative disorders, cancer, and diabetes, are the principal cause of death and disability worldwide [1]. Most of these diseases are caused by environmental factors, with the diet being one of the main contributing factors. While the consumption of highly processed foods and sugar-sweetened beverages has been associated with a higher risk of these disorders, a healthy diet including functional foods may help in reducing or even preventing several NCDs [1,2]. Thus, in recent years, the search for bioactive food compounds and their use as substitutes of pharmacological treatments has intensified. Due to their desirable impacts on human health and limited side effects, bioactive peptides have become one of the most studied food components, being usually included into functional foods and nutraceuticals [3]. Once liberated from the source protein by enzymatic hydrolysis, gastrointestinal digestion, or food processing, bioactive peptides may act on different body systems exerting different functionalities such as antihypertensive, antioxidant, opioid, antithrombotic, hypocholesterolemic, anticancer, immunomodulatory, and antimicrobial activities, among others [4]. Moreover, it has been demonstrated that some food peptides are able to exert two or more bioactivities, acting on several systems at the same time [5]. This has made it so that multifunctional peptides have been recently recognized as more useful than peptides with single activity as they influence multiple cell processes, affecting different signaling pathways simultaneously [6]. Among food sources of multifunctional peptides, milk and meat proteins are considered some of the most studied [7]. As examples, caseinophosphopeptides have been reported to exert anticariogenic, antihypertensive, immune-enhancing, antigenotoxic, and cytomodulatory effects [3]. Whey protein lactoferrin and its derived peptide, lactoferricin, are well known by their anticancer, antitumor, immunomodulatory, and antimicrobial activities [8].

The rising evidence suggests a possible common pathophysiology among NCDs, with oxidative stress and hypertension as the main contributing factors [9,10]. Oxidative stress occurs when reactive oxygen species (ROS) overload the body's defenses or when these defenses lose their capacity to react, leading to damage of essential cell components [11]. Experimental, clinical, and epidemiological studies have revealed that this status is involved in the development of NCDs such as arteriosclerosis, obesity, type 2 diabetes, inflammatory bowel disease, arthritis, neurological, liver, and renal disorders, and cancer [12]. Hypertension, defined as high blood pressure, is currently considered one of the major preventable risk factors linked to cardiovascular diseases [13].

The classical or empirical approach, also referred to as the in vitro method, is the most employed approach in peptide bioactivity screening. However, it requires exhaustive sample preparation and does not always allow the explicit identification of particular bioactive peptides. To overcome these major disadvantages, bioinformatics-driven (in silico) approaches have recently been developed. These strategies enable estimating potential precursor proteins through the calculation of quantitative descriptors, constructing profiles of the potential biological activity of peptide sequences, and predicting peptidic bonds susceptible to enzymatic hydrolysis [14]. Thus, in silico analyses have been recognized as playing a significant role in the process of bioactive peptides generation and identification [15]. However, there are many limitations associated with bioinformatic data, such as the extent, quality, and reliability of published information within databases. Moreover, the validation of attributed bioactive sequences, as well as the study of other aspects such as peptides' stability, bioavailability, and mechanisms of action should be carried out to complete the provided information by in silico analyses. Therefore, different milk and meat-derived peptides that are similar to or share amino acids with previously described opioid fragments were selected and subjected to an integrated approach combining an in silico study and in vitro assays to evaluate their multifunctionality and mechanisms of action as basis of their future use in functional foods.

2. Materials and Methods

2.1. Materials

Triisopropyl silane, angiotensin-converting enzyme (ACE), fluorescein (FL), 2,2'-azinobis(3-ethylbenzothiazoline-6-sulfonic acid) diammonium salt (ABTS), 3-(4,5-dimethylthiazol-2-yl)-2,5-diphenyl tetrazolium bromide (MTT), dimethylsulfoxide (DMSO), dichlorofluorescin (DCFH), and Hank's Balanced Salt Solution (HBSS) were purchased from Sigma-Aldrich (St. Louis, MO, USA). 2,20-azobis (2-methylpropionamide) dihydrochloride (AAPH) and 6-hydroxy-2,5,7,8-tetramethylchroman-2-carboxylic acid (Trolox) were obtained from Aldrich (Milwaukee, WI, USA). Abz-Gly-Phe (NO$_2$)-Pro was purchased from Bachem Feinchemikalien (Bubendorf, Switzerland), and trifluoroacetic acid (TFA) was obtained from Scharlau (Barcelona, Spain). High-Glucose Dulbecco's Modified Eagle Medium (DMEM), fetal bovine serum (FBS), and penicillin/streptomycin/amphotericin B solution were purchased from Biowest (Kansas City, MO, USA). A 1% non-essential amino acids solution was from Lonza Group Ltd. (Basilea, Switzerland).

The milk and meat-derived peptides used in this study (Table 1) were synthesized by the conventional Fmoc solid-phase synthesis method using an Applied Biosystems model 433A synthesizer (Foster City, CA, USA). The cleavage of the peptides from the polystyrene-based resin (Applied Biosystems, Foster City, CA, USA) was carried out with TFA, triisopropyl silane, and MilliQ water for

2 h. The analysis of peptides was carried out by high performance liquid chromatography coupled to tandem mass spectrometry (HPLC-MS/MS) analysis using an Agilent 1100 HPLC System (Agilent Technologies, Waldbron, Germany) connected online to an Esquire 3000 ion trap (Bruker Daltonik GmbH, Bremen, Germany) and equipped with an electrospray ionization source. The identification of peptides and their purity were determined through mass comparison and peak integration, respectively (Supplementary Figure S1). Finally, peptides were dissolved in 10% acetic acid, freeze-dried, and kept at −20 °C until further analysis.

2.2. Peptide Screening by In Silico Analysis

Synthetic peptides were subjected to in silico analysis by using the Milk Bioactive Peptide Database (MBPDB [16] and BIOPEP-UWM database of bioactive peptides (BIOPEP-UPW) [17]).

Table 1. Source protein, fragment, sequence, molecular mass, and purity of synthetic peptides used in the present study.

Source Protein	Sequence	Fragment	Molecular Mass (Da)	Purity (%)
α_{S1}-casein	RY	f(90–91)	337.39	99.1
	RYL	f(90–92)	450.57	99.6
	RYLG	f(90–93)	507.64	99.4
	RYLGY	f(90–94)	670.83	88.0
	RYLGYLE	f(90–96)	913.14	73.4
	YLG	f(91–93)	351.44	92.0
	YLGY	f(91–94)	514.63	97.8
	YLGYLE	f(91–96)	756.94	98.9
	LGY	f(92–94)	351.44	98.9
α_{S1}-casein	AYFYPE	f(143–148)	788.92	99.2
	YFYPEL	f(144–149)	831.01	100.0
	FYPEL	f(145–149)	667.82	99.0
β-casein A2	YPFPGPI	f(60–66)	790.02	96.0
	YPFPGPIP	f(60–67)	887.15	92.9
	YPFPGPIN	f(60–68)	904.14	94.0
β-casein	YPFVE	f(51–55)	653.79	95.0
	YPFVEP	f(51–56)	750.92	100.0
	YGFL	f(59–62)	498.63	98.5
	YGFLP	f(59–63)	595.76	100.0
	YPVEPF	f(114–119)	750.92	92.3
α-La	YGLF	f(50–53)	498.63	97.6
β-Lg	YLL	f(102–104)	407.55	100.0
	YLLF	f(102–105)	554.74	100.0
	LLF	f(103–105)	391.55	96.5
β-Hg	YPW	f(34–36)	464.55	84.5
	YPWT	f(34–37)	565.67	91.7
	PWT	f(35–37)	402.48	86.3

The potential of peptides to be bioactive was predicted using PeptideRanker software, and their theoretical bioactivity was expressed as score values calculated (from 0 to 1, with 1 being the most likely to be bioactive). Moreover, prediction of the toxicity was performed using ToxinPred.

2.3. Angiotensin Converting Enzyme (ACE)-Inhibitory Activity

The ACE-inhibitory activity of synthetic peptides was determined by the fluorescence protocol optimized by Sentandreu and Toldrá (2006) [18] and modified by Quirós et al. (2009) [19]. Briefly, the substrate Abz–Gly–Phe(NO$_2$)–Pro was dissolved (0.45 mM) in 150 mM Tris and 1125 mM NaCl buffer (pH 8.3) and maintained at 4 °C until its use. ACE (1 U/mL) was diluted (0.04 U/mL) in 150 mM Tris buffer containing 0.1 μM ZnCl$_2$ (pH 8.3). Forty microliters of sample (or MilliQ water for blank and control) were added to a black multi-well plate (Porvair, Leatherhead, UK). Then, 40 μL of ACE were added, and the reaction started after the addition of 160 μL of the substrate. The plate was incubated at 37 °C for 30 min, and the fluorescence was measured in a FLUOstar OPTIMA plate reader (BMG Labtechnologies GmbH, Offenburg, Germany) with 320 nm excitation and 420 nm emission filters. Data were processed with the FLUOstar Control version 1.32 R2 (BMG Labtech) software and expressed as IC_{50} (peptide concentration needed to inhibit 50% of the ACE activity).

2.4. In Vitro Antioxidant Activity

2.4.1. Oxygen Radical Absorbance Capacity (ORAC)-FL Assay

An oxygen radical absorbance capacity (ORAC)-FL assay was used based on the protocol previously optimized [20]. The reaction was performed at 37 °C in 75 mM phosphate buffer (pH 7.4). The final assay mixture volume was 200 μL, containing 70 nM FL, 12 mM AAPH, and antioxidant [(Trolox, 1–8 μM) or sample (at different concentrations)]. Fluorescence was measured during 137 min in a FLUOstar OPTIMA plate reader (BMG Labtech) with 485 nm excitation and 520 nm emission filters. The equipment was controlled by the FLUOstar Control ver. 1.32 R2 software for fluorescence measurement. Samples were analyzed in triplicate. The final ORAC-FL value was expressed as μmol Trolox equivalents (TE) per μmol peptide.

2.4.2. ABTS Assay

Antioxidant activity was measured using a previously optimized method [21] with some modifications. A mixture of 7 mM ABTS stock solution and 2.45 mM potassium persulfate was kept in the dark at room temperature for 12-16 h to form the ABTS radical cation ($ABTS^{\bullet+}$). The $ABTS^{\bullet+}$ solution was diluted in 5 mM of phosphate buffer solution (PBS) (pH 7.4) to an absorbance of 0.70 ± 0.02 at 734 nm at 30 °C. Two mL of diluted $ABTS^{\bullet+}$ solution were mixed with 20 μL of sample or Trolox (0–0.015 μmol), and the absorbance was recorded at 734 nm after 10 min incubation at 30 °C. The Trolox equivalent antioxidant capacity (TEAC) value was expressed as μmol TE per μmol peptide. Each sample was analyzed in triplicate.

2.5. Cell Culture

The human colorectal adenocarcinoma Caco-2 and the mouse macrophage RAW 264.7 cell lines were obtained from the American Type Culture Collection (ATCC, Rockville, MD, USA). Caco-2 and RAW264.7 cells were grown in High-Glucose DMEM supplemented with 10% (v:v) FBS, 1% (v:v) penicillin/streptomycin/amphotericin B, and 1% (v:v) non-essential amino acids solution. Cells were maintained in plastic 75-cm^2 culture flasks at 37 °C in a humidified incubator containing 5% CO_2 and 95% air.

2.6. Cell Treatment Conditions

Cells were incubated for 24 h with various concentrations of selected synthetic peptides. To evaluate both the direct and protective effects against oxidative stress, the incubation period was followed by 1.5 h-treatment with culture medium or *tert*-butyl hydroperoxide (*t*-BOOH, 1 mM for Caco-2 cells and 0.25 mM for RAW264.7 cells), respectively, and different biomarkers were evaluated.

2.6.1. Cell Viability

Cell viability was determined using the MTT assay. Caco-2 and RAW264.7 cells were seeded onto 96-well plates (Corning Costar Corp., Corning, NY, USA) at a density of 1.0×10^4 cells/cm^2 and incubated for 9 days and 24 h, respectively. Then, cells were washed with PBS, treated with synthetic selected peptides (1–100 μM), and incubated for 24 h. Afterwards, culture medium was removed, and the cells were washed with PBS and incubated with medium (direct effects) or chemical oxidant (protective effects) at 37 °C for 1.5 h. At the end of the treatment time, 100 μL of MTT solution (0.5 mg/mL final concentration) were added to each well, and the plate was incubated for 2 h at 37 °C. The supernatant was aspirated, the formazan crystals were solubilized in DMSO:ethanol (1:1, v:v), and the absorbance was measured at 570 nm in a FLUOstar OPTIMA plate reader (BMG Labtech). Results were expressed as percentage of the control, considered as 100%. Samples were analyzed in triplicate.

2.6.2. Determination of Intracellular Reactive Oxygen Species (ROS)

The intracellular ROS levels were detected using the ROS-sensitive fluorescent dye, DCFH, as previously described [22]. Caco-2 cells were plated in 48-well plates (density of 4.75×10^4 cells/well) and RAW264.7 cells in 24-well plates (density of 2.0×10^5 cells/well) and incubated for 7 days and 14 h, respectively. After this time, cells were treated with peptides as previously described for 24 h. Then, DCFH was dissolved in HBSS and added to cells at a final concentration of 5 µM solution. Cells were incubated for 30 min at 37 °C. The probe was removed, and cells were incubated with PBS (direct effects) or *t*-BOOH (protective effects) for 60 min, measuring the fluorescence after 60 min in a FLUOstar OPTIMA plate reader (BMG Labtech) with 485 nm excitation and 520 nm emission filters. The results were expressed as percentage of the control, which was considered as 100%. The assay was run in triplicate.

2.7. Statistical Analyses

All data were analyzed from three independent experiments. Results were expressed as the mean ± standard deviation (SD). Data were statistically analyzed by performing a one-way ANOVA test, followed by Tukey's multiple comparison test with the IBM SPSS Statistics for Windows 23.0 (IBM Corporation, Armonk, NY, USA). A *p*-value of less than 0.05 was considered statistically significant. Significant differences of each concentration versus the control under the same experimental conditions were expressed by *** ($p < 0.001$), ** ($p < 0.01$), and * ($p < 0.05$).

3. Results and Discussion

3.1. In Silico Analysis of Synthetic Peptides

The physicochemical characteristics and predicted toxicity and biological activity of synthetic peptides are shown in Table 2. The bioactivity of peptides was predicted using the PeptideRanker program. There were differences in the theoretical bioactivity of peptides, with score values from 0.2453 to 0.9558. Eleven peptides, released from β-casein (β-CN), β-lactoglobulin (β-Lg), α-lactalbumin (α-La), and β-hemoglobin (β-Hg), were to be highly bioactive with a predicted bioactive score over 0.80. None of the 27 peptides were considered toxic, according to the software ToxinPred. The results of the in silico analysis by using the MBPDB and BIOPEP-UWM databases are shown in Table 3. Of the 27 synthetic peptides studied, 23 were already included into databases because of their opioid, ACE-inhibitory, antioxidant, anticancer, antidiabetic, or immunostimulating activities, among others. Some of them were found to exert two or more activities. Thus, α_{s1}-CN-derived peptides RY, RYL, RYLG, YLGY and FYPEL exert both ACE-inhibitory and antioxidant activities [23–27] ACE-inhibitory and opioid activities are exerted by sequences YGFLP [28] and YGFL [29,30]. Four of the analyzed peptides (RYLGY, LGY, YPFPGPI, and YLLF) have been reported to exert three or more activities. Among them, the multifunctional β-casein A2-derived peptide YPFPGPI is highlighted by exerting ACE and dipeptidyl peptidase IV (DPP-IV) inhibitory, anticancer, anxiolytic, immunomodulatory, opioid, antidiabetic, and satiating activities (Table 3). Only four peptides (YPFPGPIP, YPFVEP, YGFL, and YPW) were found to be novel, although they share active sequences with bioactive peptides released from the same source protein.

Table 2. Physicochemical characteristics and predicted biological activity and toxicity of synthetic peptides derived from food sources (PeptideRanker and ToxinPred databases).

Peptide	Hydrophobicity	Hydrophilicity	Charge	pI [1]	Toxicity Prediction	Activity Prediction
RY	−0.87	0.35	1.00	9.10	Non toxin	0.5437
RYL	−0.40	−0.37	1.00	9.10	Non toxin	0.5627
RYLG	−0.26	−0.27	1.00	9.10	Non toxin	0.5215
RYLGY	−0.21	−0.68	1.00	8.93	Non toxin	0.4505
RYLGYLE	−0.16	−0.31	0.00	6.35	Non toxin	0.2453
YLG	0.24	−1.37	0.00	5.88	Non toxin	0.6404
YLGY	0.18	−1.60	0.00	5.87	Non toxin	0.6138
YLGYLE	0.11	−0.87	−1.00	4.00	Non toxin	0.3219
LGY	0.24	−1.37	0.00	5.88	Non toxin	0.5959
AYFYPE	0.04	−0.77	−1.00	4.00	Non toxin	0.7126
YFYPEL	0.08	−0.98	−1.00	4.00	Non toxin	0.7603
FYPEL	0.09	−0.72	−1.00	4.00	Non toxin	0.7939
YPFPGPI	0.19	−0.94	0.00	5.88	Non toxin	0.9175
YPFPGPIP	0.15	−0.82	0.00	5.88	Non toxin	0.8990
YPFPGPIPN	0.08	−0.80	0.00	5.88	Non toxin	0.8061
YPFVE	0.10	−0.66	−1.00	4.00	Non toxin	0.4339
YPFVEP	0.07	−0.55	−1.00	4.00	Non toxin	0.5114
YGFL	0.33	−1.65	0.00	5.88	Non toxin	0.9558
YGFLP	0.25	−1.32	0.00	5.88	Non toxin	0.9432
YPVEPF	0.07	−0.55	−1.00	4.00	Non toxin	0.6345
YGLF	0.33	−1.65	0.00	5.88	Non toxin	0.9537
YLL	0.36	−1.97	0.00	5.88	Non toxin	0.6000
YLLF	0.42	−2.10	0.00	5.88	Non toxin	0.9038
LLF	0.56	−2.03	0.00	5.88	Non toxin	0.9389
YPW	0.11	−1.90	0.00	5.88	Non toxin	0.9751
YPWT	0.04	−1.52	0.00	5.88	Non toxin	0.8795
PWT	0.04	−1.27	0.00	5.88	Non toxin	0.8928

[1] pI: Isoelectric point.

Table 3. Predicted biological activity of synthetic peptides derived from food sources using Milk Bioactive Peptide Database (MBPDB) and BIOPEP-UWM database of bioactive peptides.

Sequence	Biological Activity	Results	Reference
RY	ACE inhibitory Antioxidant	IC_{50} [a] = 51.00 μM */54.43 μM ** ORAC = 1.94 μmol TE/μmol peptide **	[23] */[24] ** [24] **
RYL	ACE inhibitory Antioxidant	IC_{50} [a] = 3.31 μM */106.64 μM ** ORAC = 1.75 μmol TE/μmol peptide **	[25] */[26] ** [24] **
RYLG	ACE inhibitory Antioxidant	IC_{50} [a] = 224.69 μM ** ORAC = 1.67 μmol TE/μmol peptide **	[24] ** [24] **
RYLGY	ACE inhibitory Antioxidant Opioid	IC_{50} [a] = 0.71 μM *,** ORAC = 2.83 μmol TE/μmol peptide ** Stimulation of mucin secretion **	[26] *,** [24] ** [31] **
RYLGYLE	Opioid Anticancer	IC_{50} [b] = 1.2 μM * Decrease of breast cancer cell proliferation **	[32] * [33] **
YLG	Antioxidant	ORAC = 1.38 μmol TE/μmol peptide **	[24] **
YLGY	ACE inhibitory Antioxidant	IC_{50} [a] = 41.86 μM *,** ORAC = 1.46 μmol TE/μmol peptide **	[24] *,** [26] **
YLGYLE	Opioid	IC_{50} [b] = 45.00 μM * Stimulation of mucin secretion **	[32] * [31] **
LGY	Immunostimulating ACE inhibitory Antioxidant	n.d. IC_{50} [a] = 21.46 μM ** ORAC = 2.31 μmol TE/μmol peptide **	[34] * [24] ** [24] **
AYFYPE	ACE inhibitory	IC_{50} [a] = 106.00 μM *,**/260.82 μM **	[35] *,**/[24] **
YFYPEL	Antioxidant Opioid	DPPH value = 79.20 μM ** Increase MUC5AC expression	[27] ** [36,37] **
FYPEL	ACE inhibitory Antioxidant	IC_{50} [a] = 80.60 μM ** ORAC = 1.77 μmol TE/μmol peptide **/DPPH = 127.50 μM **	[24] ** [24] **/[27] **

Table 3. *Cont.*

Sequence	Biological Activity	Results	Reference
YPFPGPI	ACE inhibitory	IC_{50} [a] = 500.00 μM **	[38] **
	Anticancer	Decrease of breast cancer cell proliferation **	[33] **
	Anxiolytic	Induction of inflammatory immune response in gut **	[39] **
	Immunomodulatory	Inhibition/stimulation of lymphocyte proliferation at low/high concentrations **	[40] **
	Opioid	Stimulation of lymphocyte proliferation [d] = −21/+26 **	[41] **
		IC_{50} [c] = 14 μM **	[42] **
		Increase of jejunal mucus secretion and mucus discharge ** Increase of MUC2 and MUC3 expression in DHE cells **	[43] **
		Increase of MUC5A expression in HT29-MTX cells **	[44] **
		Stimulation of mucin secretion **	[36,45] **
	Antidiabetic	Reduction of pancreas MDA level in diabetic rats **	[46] **
	Satiating	Induction of CCK-8 **	[47] **
YPFPGPIP	n.d.	n.d	n.d
YPFPGPIPN	ACE inhibitory	IC_{50} [a] = 14.80 μM **	[48] **
	Antidiabetic	IC_{50} [e] = 6.70 μM **	[49] **
YPFVE	Opioid	Stimulation of mucin secretion **	[50] **
YPFVEP	n.d.	n.d	n.d
YGFL	n.d.	n.d	n.d
YGFLP	ACE inhibitory	IC_{50} [a] = 260.00 μM *	[28] *
	Opioid agonist	n.d.	n.d.
YPVEPF	Antidiabetic	IC_{50} [e] = 124.70 μM *	[51] *
	Opioid	IC_{50} [c] = 59.00 μM **	[52] **
		Increase of MUC4 expression **	[53] **
YGLF	ACE inhibitory	IC_{50} [a] = 733.30 μM *	[30] *
	Opioid agonist	IC_{50} [c] = 300.00 μM **	[29] **
YLL	Antioxidant	FRAP = 81.76 mmol Fe/mol peptide **	[54] **
YLLF	ACE inhibitory	IC_{50} [a] = 171.80 μM *	[30] *
	Opioid agonist	IC_{50} [c] = 160.00 μM *	[29] *
		Stimulation of mucin secretion **	[36,50,55] **
	Cytotoxic	Stimulation of murine splenocytes **	[56] **
LLF	ACE inhibitory	IC_{50} [a] = 79.80 μM *	[57] *
YPW	n.d.	n.d.	n.d.
YPWT	Opioid	IC_{50} [c] = 45.20 μM *	[58] *
PWT	Antioxidant	Inhibition of linoleic acid peroxidation *	[48] *

* According to BIOPEP-UWM database; ** According to Milk Bioactive Peptide Database (MBPDB); IC_{50} [a]: Values (μM) are given for peptide concentrations inhibiting the angiotensin-converting enzyme (ACE) activity by 50%; IC_{50} [b]: Values (μM) is given for peptide concentration inhibiting (3H)-dihydromorphine binding, instead of (3H)-naloxone, by 50%; IC_{50} [c]: Values (μM) are given for peptide concentrations inhibiting (3H)-naloxone binding by 50%; Stimulated lymphocyte proliferation [d]: % stimulation (+) and inhibition (−), respectively, compared to control; IC_{50} [e]: Values (μM) are given for peptide concentration required to inhibit 50% of dipeptidyl peptidase IV (DPP-IV); n.d. No available data. ORAC: oxygen radical absorbance capacity.

3.2. In Vitro ACE Inhibitory and Antioxidant Activities of Synthetic Peptides

The measured ACE-inhibitory and antioxidant activities of assayed peptides are shown in Table 4. Our study confirmed the ACE-inhibitory activity already reported for several of the analyzed fragments. Moreover, this effect was newly found in other sequences such as YLGYLE, YFYPE, YPFPGPIPN, YPFVEP, YGFL, and YLL. Potent activity (IC_{50} values lower than 10 μM) was determined for α_{s1}-CN peptide YFYPEL (IC_{50} = 8.82 μM) and β-CN peptide YPFVEP (IC_{50} = 7.48 μM). These values were similar to those reported for well-known milk derived tripeptides VPP (IC_{50} = 9 μM) and IPP (IC_{50} = 5 μM) [59]. The presence of leucine, valine, and proline at the C-terminus could contribute to the high ACE-inhibitory activity shown by these two sequences, as ACE prefers substrates/competitive inhibitors with hydrophobic amino acids and/or proline at the three C-terminal positions [60]. Similarly, milk peptide HLPLP, containing leucine and proline at the C-terminus, showed an IC_{50} value of 41 µM [28]. As shown in Table 4, all peptides but sequence LLF presented antioxidant activity through a dual mechanism of action, hydrogen atom transfer (ORAC), and non-competitive electron transfer (ABTS). Our results confirmed the peroxyl radical scavenging activity already reported for eight of these peptides. Moreover, this property was newly found in 18 peptides whose ORAC values ranged from 1.09 ± 0.02 µmol TE/µmol peptide to 3.50 ± 0.02 µmol TE/µmol peptide. The highest value was determined for peptide YPW. The presence of tyrosine and tryptophan could determine its potent activity, as these two amino acids have been reported to be the main contributors to the peroxyl radical scavenging activity of food-derived peptides [20]. Moreover, the situation of these residues at the peptide chain could also influence the activity. Thus, when threonine was added to the C-terminus of peptide YPW, the ORAC value was reduced up to 3.19 ± 0.18 µmol TE/µmol peptide. This influence was also observed for peptides YLG and LGY. For peptide LG, the antioxidant behavior of the resultant peptides from the addition of tyrosine as a terminal residue was different. Thus, if this amino acid (Y) was added to the C-terminus, the ORAC value of peptide LGY was 2.00 ± 0.09 µmol TE/µmol peptide, while it was 0.93 ± 0.08 µmol TE/µmol peptide when tyrosine was added to the N-terminus (peptide YLG). When tyrosine was added at both termini, the ORAC value increased up to 2.96 ± 0.20 µmol TE/µmol peptide (Table 4). Our results confirmed the previously described importance of peptides parameters such as the their amino acid composition, sequence, and length in determining their antioxidative potential [61].

Table 4. ACE-inhibitory activity (expressed as µM) and antioxidant activity (expressed as µmol Trolox equivalents (TE)/µmol peptide of synthetic animal-protein derived peptides.

Sequence	ACE [1] Inhibitory Activity (IC_{50}-µM)	Antioxidant Activity (µmol TE[2]/µmol Peptide)	
		ORAC	TEAC
RY	*	1.83 ± 0.13	1.38 ± 0.03
RYL	*	1.72 ± 0.14	1.90 ± 0.03
RYLG	*	1.70 ± 0.11	2.91 ± 0.21
RYLGY	3.08 ± 0.11	2.97 ± 0.09	1.38 ± 0.14
RYLGYLE	*	2.88 ± 0.07	3.10 ± 0.01
YLG	*	0.93 ± 0.08	1.40 ± 0.04
YLGY	9.87 ± 0.31	2.96 ± 0.20	2.14 ± 0.11
YLGYLE	85.76 ± 4.66	2.28 ± 0.22	5.96 ± 0.35
LGY	26.10 ± 0.83	2.00 ± 0.09	1.54 ± 0.01
AYFYPE	774.36 ± 38.22	2.60 ± 0.13	1.99 ± 0.19
YFYPEL	8.82 ± 0.58	2.66 ± 0.16	2.59 ± 0.17
FYPEL	62.00 ± 6.27	1.88 ± 0.13	1.74 ± 0.04
YPFPGPI	685.91 ± 102.91	1.91 ± 0.14	1.62 ± 0.08
YPFPGPIP	224.05 ± 43.94	1.09 ± 0.02	1.86 ± 0.01
YPFPGPIN	378.65 ± 11.15	1.26 ± 0.06	1.22 ± 0.17
YPFVE	*	1.53 ± 0.15	1.78 ± 0.02
YPFVEP	7.48 ± 0.03	1.96 ± 0.14	1.43 ± 0.08
YGFL	292.53 ± 0.83	1.42 ± 0.04	2.12 ± 0.03
YGFLP	272.39 ± 0.25	2.27 ± 0.15	2.22 ± 0.02
YPVEPF	*	1.62 ± 0.09	1.75 ± 0.10
YGLF	*	0.89 ± 0.01	2.08 ± 0.06
YLL	518.54 ± 3.50	0.78 ± 0.03	2.55 ± 0.29
YLLF	n.d.	0.91 ± 0.04	1.96 ± 0.30
LLF	94.79 ± 2.97	**	**
YPW	*	3.50 ± 0.02	2.32 ± 0.09
YPWT	*	3.19 ± 0.18	3.89 ± 0.10
PWT	*	2.15 ± 0.07	0.73 ± 0.07

[1] ACE: Angiotensin-converting enzyme; [2] TE: Trolox equivalents; TEAC: Trolox equivalent antioxidant capacity; n.d. Activity not determined. * ACE-inhibitory activity not detected at the highest peptide concentration analyzed (1000 µM). ** Antioxidant activity not detected at the highest peptide concentration analyzed (0.20 µmol).

Unlike peroxyl radical scavenging activity, none of the analyzed peptides had been previously reported to exert ABTS radical scavenging properties. TEAC values ranged from 0.73 ± 0.07 µmol

TE/µmol peptide to 5.96 ± 0.35 µmol TE/µmol peptide (Table 4). As it has been described for the ORAC assay, the presence of tyrosine and tryptophane was responsible for the ABTS radical scavenging properties of analyzed peptides [21]. Among peptides whose antioxidant activity has been described for the first time in the present study, six sequences showed potent effects with ORAC and TEAC values higher than 2.0 µmol TE/µmol peptide. These sequences corresponded to α_{s1}-CN peptides RYLGYLE, YLGYLE, and YFYPEL, β-CN peptide YGFLP, and β-Hg peptides YPW and YPWT. Five of these peptides were known by their opioid activity [31,32,36,37,42]. Additionally, anticancer and ACE-inhibitory activities have been reported for peptides RYLGYLE [33] and YGFLP [28], respectively. Thus, the antioxidant activity described in the present study would increase the functionality of these food-derived peptides. As a result of their multifunctionality allowing peptides to exert beneficial effects on different body systems, sequences RYLGYLE, YLGYLE, YFYPEL, and YPWT were selected to study in depth the mechanism of action involved in their antioxidant activity.

3.3. Antioxidant Activity of Synthetic Peptides in Cell Models

Two cell models, human colon adenocarcinoma Caco-2 and murine macrophages RAW264.7, were used to evaluate the protective effects of animal protein-derived peptides on the cell oxidative status under normal and chemical-induced conditions. The action of peptides on cell viability and ROS generation was studied. The direct effects of peptides RYLGYLE, YLGYLE, YFYPEL, and YPWT on Caco-2 and RAW264.7 cells viability were evaluated using the MTT assay. This assay provides a sensitive measurement of the metabolic status of the mitochondria, which reflects early cellular redox changes [62]. Treatment of Caco-2 and RAW264.7 cells with synthetic peptides did not evoke changes in cell viability, indicating that the concentrations selected (1–100 µM) did not damage cell integrity during the 24-h period of incubation.

To study the protective effects of peptides against chemical-induced oxidative damage in Caco-2 cells and macrophages, they were pre-incubated with peptides for 24 h, exposed to *t*-BOOH for 1.5 h, and then, cell viability was measured. As shown in Figure 1A–D, treatment of Caco-2 cells with *t*-BOOH (1 mM) provoked a significant reduction of cell viability of 25%, compared to non-stimulated cells.

In a previous study in our lab, García-Nebot et al. (2014) [22] had reported a 20% reduction of the Caco-2 cells' viability after 1.5-h treatment with 3 mM *t*-BOOH. A longer incubation time (6 h) with this chemical at concentrations of 0.1 and 4 mM has been also found to cause significant reductions of cell viability [63,64]. Similarly, the viability of macrophages was significantly reduced after treatment with *t*-BOOH (0.25 mM) by up to 44% (Figure 2A–D). Recent studies have also reported significant decreases (39%) of cell viability after treatment of RAW264.7 cells with 1 mM *t*-BOOH for 3 h [65]. Pre-treatment of Caco-2 cells with tested peptides before induction with *t*-BOOH for 1.5 h did not exert any protection from the effects of this chemical. In the case of macrophages, only treatment with peptide YPWT resulted in an increase of cell viability at concentrations between 25 and 100 µM. Thus, the percentage of viable cells increased from 70.97% (stimulated cells) to 79.58% (stimulated cells pre-treated with 25 µM YPWT) (Figure 2D). These results are in agreement with previous studies carried out with soybean protein-derived peptides that only found significant protection on *t*-BOOH-induced RAW264.7 cells at high doses [66].

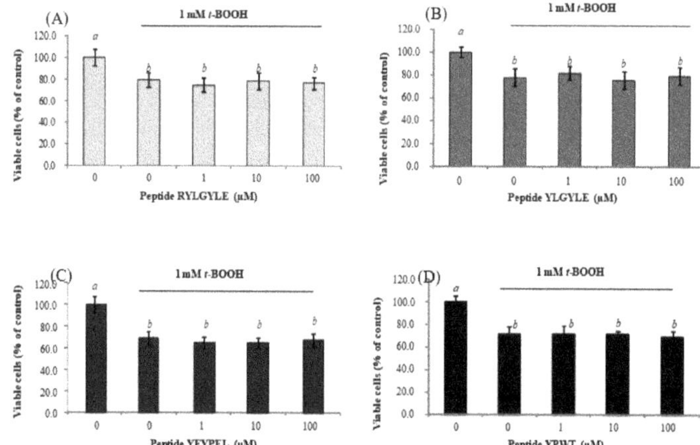

Figure 1. Dose-dependent effects of synthetic animal-protein derived peptides (**A**) RYLGYLE, (**B**) YLGYLE, (**C**) YFYPEL, and (**D**) YPWT on cell viability of stressed Caco-2 cells with 1 mM tert-butyl hydroperoxide (*t*-BOOH). Cells were pre-treated with peptides at concentrations ranged from 1 to 100 µM for 24 h. Results were expressed as the percentage of viable cells compared to control, which was considered as 100% (% control, mean ± standard deviation (SD), n = 3). Different letters indicate significant differences ($p < 0.05$; Tukey multiple comparison test).

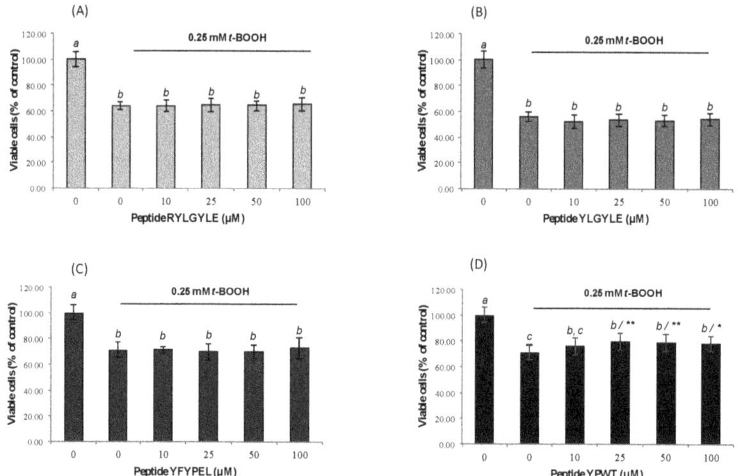

Figure 2. Dose-dependent effects of synthetic animal protein-derived peptides (**A**) RYLGYLE, (**B**) YLGYLE, (**C**) YFYPEL, and (**D**) YPWT on cell viability of stressed macrophages RAW264.7 with 0.25 mM tert-butyl hydroperoxide (*t*-BOOH). Cells were pre-treated with peptides at concentrations that ranged from 10 to 100 µM for 24 h. Results were expressed as the percentage of viable cells compared to control, considered as 100% (% control, mean ± standard deviation (SD), n = 3). Different letters indicate significant differences ($p < 0.05$) and ** ($p < 0.01$); * ($p < 0.05$) significant differences of each concentration versus control under the same experimental conditions (one-way ANOVA followed by Tukey's multiple comparison test).

In order to understand the potential mechanism of cytoprotective action exerted by peptides, the intracellular ROS generation was evaluated in normal cells and cells exposed to *t*-BOOH after pre-treatment with synthetic peptides for 24 h. The direct evaluation of intracellular ROS is recognized as a good indicator of the oxidative damage to living cells [67]. In our study, measurement of the intracellular ROS levels was carried out using DCFH as a fluorescent probe that once added to intact cells, crosses cell membranes and is oxidized to highly fluorescent dichlorofluorescein (DCF) in the presence of ROS [68]. In Caco-2 cells under normal conditions, four peptides caused a significant reduction of ROS levels at all concentrations used (Figure 3A,C,E,G).

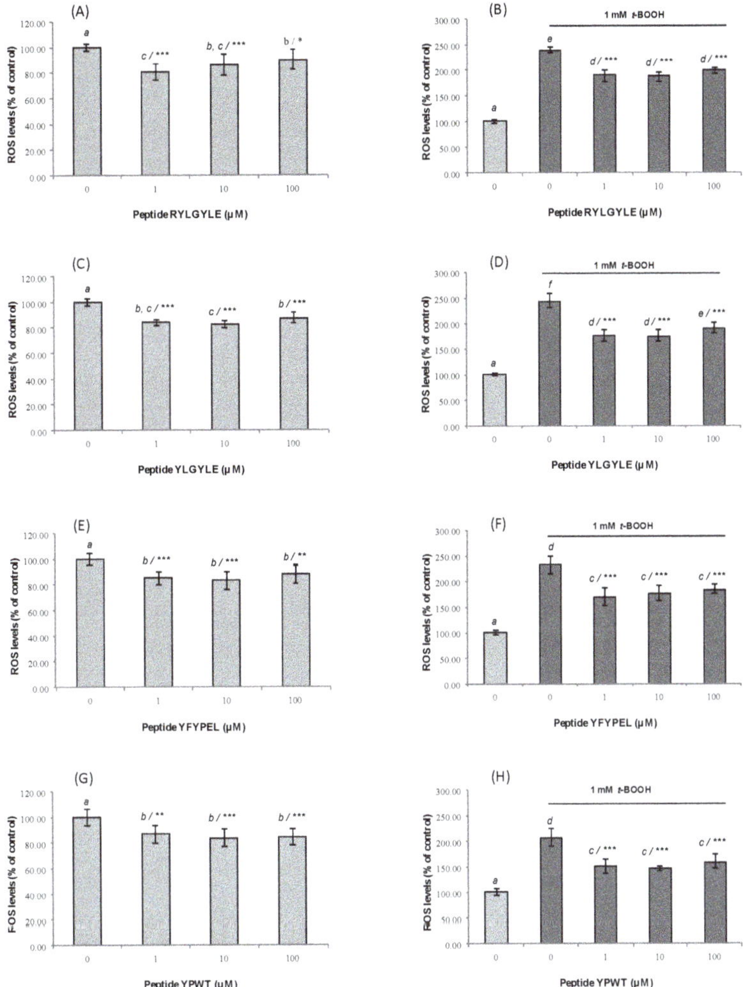

Figure 3. Dose-dependent effects of synthetic animal protein-derived peptides (**A,B**) RYLGYLE, (**C,D**) YLGYLE, (**E,F**) YFYPEL, and (**G,H**) YPWT on reactive oxygen species (ROS) production in non-stressed Caco-2 cells (**A,C,E,F**) and Caco-2 cells stressed with tert-butyl hydroperoxide (*t*-BOOH, 1 mM). Cells were pre-treated with peptides at concentrations that ranged from 1 to 100 µM for 24 h. Results were expressed as the percentage of ROS levels compared to control, which were considered as 100% (% control, mean ± standard deviation (SD), n = 3. Different letters indicate significant differences ($p < 0.05$) and *** ($p < 0.001$); ** ($p < 0.01$); * ($p < 0.05$) indicate significant differences of each concentration versus control under the same experimental conditions (one-way ANOVA followed by Tukey's multiple comparison test).

In the case of fragments RYLGYLE and YLGYLE, the highest reduction (19.3% and 17.3%) was observed after treating cells with low concentrations of these peptides (1 and 10 μM, respectively), compared to non-treated cells. However, these two peptides at 10 μM provoked an increase of ROS levels in non-stressed RAW264.7 (Figure 4A,C) cells. These results indicated that in addition to the peptide dose, the type of cell line could influence the antioxidant activity of the peptides.

The chemical exposition of Caco-2 cells to t-BOOH for 1.5 h significantly increases ROS levels (untreated Caco-2 cells 100.00 ± 4.15%; treated Caco-2 cells with 1 mM t-BOOH 230.19 ± 13.61%) ($p < 0.05$) (Figure 3B,D,F,H). The pre-treatment with animal protein-derived peptides at all assayed doses for 24 h significantly neutralized the ROS-generating ability of the chemical, but no dose-dependence was observed. The highest reduction (≈68.5%) was observed for cells treated with peptide YLGYLE at 1 and 10 μM, which had ROS levels of 175.06 ± 11.13% and 174.44 ± 11.60%, respectively, compared to non-peptide-treated cells (Figure 3D). In the case of macrophages, the exposition to *t*-BOOH resulted in a higher increase of ROS levels (untreated RAW264.7 cells 100.00 ± 4.77%; treated RAW264.7 cells with 0.25 mM *t*-BOOH 501.27 ± 44.20%) (Figure 4B,D,F,H). This result was similar to that found previously by Indiano-Romacho et al. (2019) [66]. Tested peptides did not show any antioxidant activity in macrophages. No significant ROS level decreases versus control were observed after treatment with peptides both in untreated and treated RAW264.7 cells with *t*-BOOH. This tendency is in disagreement with that observed both in the Caco-2 cell model and in the in vitro antioxidant activity measured by ORAC-FL and ABTS assays; besides that, cell viability is not compromised because of the MTT results. Among cases that can help appreciate these differences, fragment YFYPEL shows one of the most remarkable activities in the stressed Caco-2 cell model, significantly lowering ROS levels (from 231.58 ± 18.40% up to 168.88 ± 17.80% at 1 μM) and showing significant in vitro antioxidant activity (2.66 ± 0.16 and 2.59 ± 0.17 μmol TE/μmol peptide for ORAC and TEAC, respectively), but without any antioxidant activity in homologous conditions in a RAW264.7 cell model. It would be interesting to evaluate the suitability of this cell model for the study of antioxidant capacity.

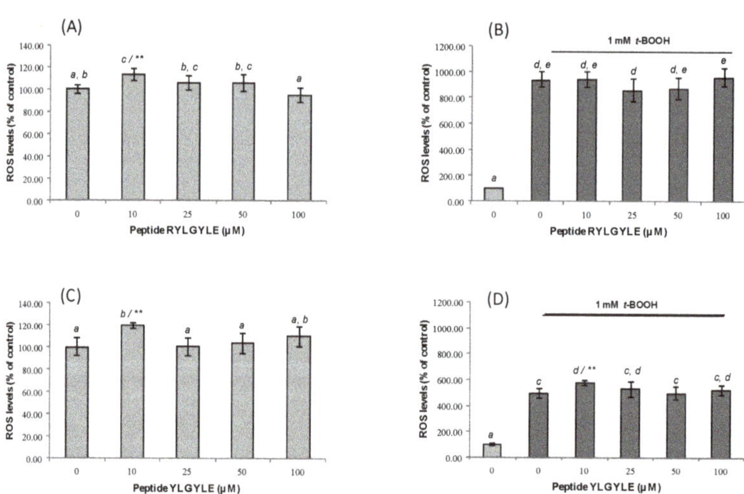

Figure 4. *Cont.*

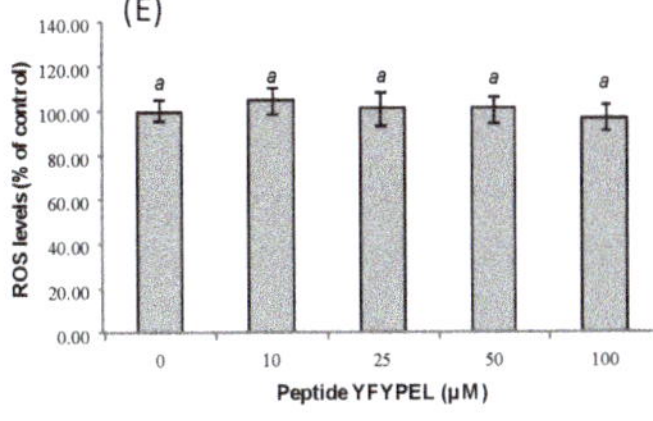
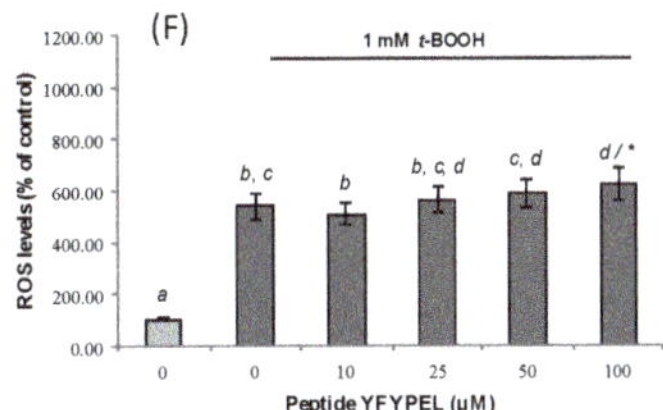
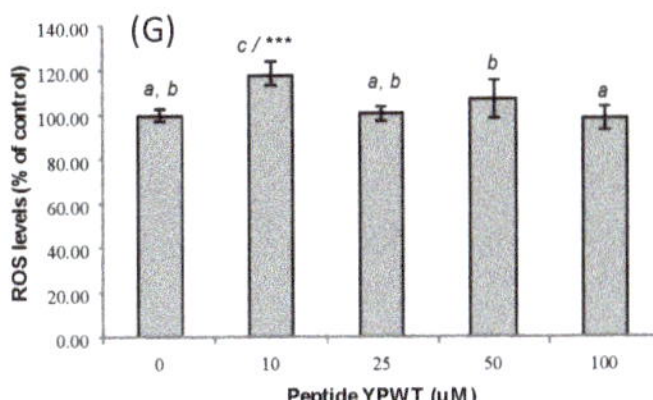
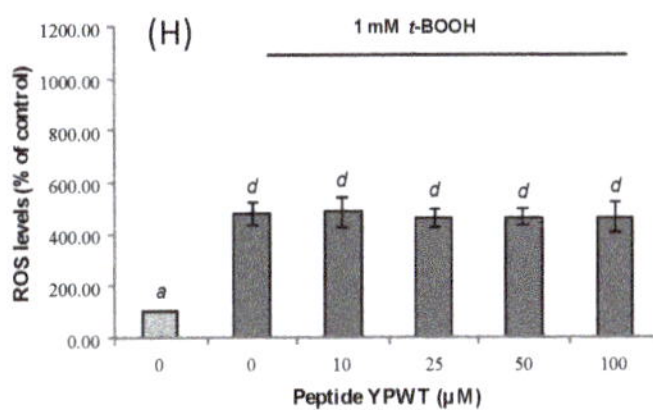

Figure 4. Dose-dependent effects of synthetic animal-protein derived peptides (**A,B**) RYLGYLE, (**C,D**) YLGYLE, (**E,F**) YFYPEL, and (**G,H**) YPWT on ROS production in non-stressed macrophages RAW264.7 (**A,C,E,F**) and macrophages RAW264.7 stressed with tert-butyl hydroperoxide (*t*-BOOH, 0.25 mM). Cells were pre-treated with peptides at concentrations that ranged from 10 to 100 μM for 24 h. Results were expressed as percentage of ROS levels compared to control, which was considered as 100% (% control, mean ± standard deviation (SD), n = 3. Different letters indicate significant differences ($p < 0.05$) and *** ($p < 0.001$); ** ($p < 0.01$); * ($p < 0.05$) significant differences of each concentration versus control under the same experimental conditions (one-way ANOVA followed by Tukey's multiple comparison test).

4. Conclusions

In summary, our results have demonstrated the multifunctionality of different bioactive peptides by an approach combining in silico and in vitro assays. By the in silico analysis, different activities such as ACE inhibitory, antioxidant, opioid, and anticancer, among others, were found to be exerted by most of analyzed peptides.

Moreover, four novel peptides (YPFPGPIP, YPFVEP, YGFL, and YPW) had not been previously defined as bioactive peptides. The ACE-inhibitory and the antioxidant activities of peptides mediated through a dual mechanism of action were confirmed by in vitro assays. Four of these peptides, RYLGYLE, YLGYLE, YFYPEL, and YPWT, were selected by their potent activity, which was confirmed in the gut epithelial cell model Caco-2, protecting cells from the oxidative damage caused by chemical agents. Future animal models would be required to confirm the multifunctionality of food-derived peptides and their protective capacity against oxidative stress-associated diseases.

Supplementary Materials: The following are available online at http://www.mdpi.com/2304-8158/9/8/991/s1, Figure S1: RP-HPLC-MS/MS analysis of synthesized peptide YPFVE at concentration of 0.2 mg/mL (A) UV chromatogram obtained at wavelength of 214 nm; (B) Peak integration of UV chromatogram; (C) Total ion current chromatogram; (D) Average mass spectrum; (E) MS/MS spectrum of selected ion m/z 654.3 (retention time 38.3 min); (F) MS/MS fragment ions identification in sequence YPFVE.

Author Contributions: Conceptualization, L.A. and B.H.-L.; methodology, D.M.-M.; investigation, D.M.-M.; writing—original draft preparation, L.A., D.M.-M., and B.H.-L.; writing-review and editing, L.A.; D.M.-M.; funding acquisition, L.A., B.H.-L. All authors have read and agreed to the published version of the manuscript.

Funding: This research was funded by project AGL2015- 66886-R from the Spanish Ministry of Science and Innovation (MICIU).

Conflicts of Interest: The authors declare no conflict of interest.

References

1. Olivieri, C. The current state of heart disease: Statins, cholesterol, fat and sugar. *Int. J. Evid. Based Healthc.* **2019**, *17*, 1. [CrossRef]
2. Ghaedi, E.; Mohammadi, M.; Mohammadi, H.; Ramezani-Jolfaie, N.; Malekzadeh, J.; Hosseinzadeh, M.; Salehi-Abargouei, A. Effects of a paleolithic diet on cardiovascular disease risk factors: A systematic review and meta-analysis of randomized controlled trials. *Adv. Nutr.* **2019**, *10*, 634–646. [CrossRef] [PubMed]
3. Agyei, D.; Potumarthi, R.; Danquah, M.K. Food-derived multifunctional bioactive proteins and peptides: Applications and recent advances. In *Biotechnology of Bioactive Compounds: Sources and Applications*; Gupta, V.K., Tuohy, M.G., Lohani, M., O'Donovan, A., Eds.; John Wiley & Sons, Ltd.: Hoboken, NJ, USA, 2015; Chapter 20.
4. Agyei, D.; Potumarthi, R.; Danquah, M.K. Food-derived multifunctional bioactive proteins and peptides: Sources and production. In *Biotechnology of Bioactive compounds: Sources and Applications*; Gupta, V.K., Tuohy, M.G., Lohani, M., O'Donovan, A., Eds.; John Wiley & Sons, Ltd.: Hoboken, NJ, USA, 2015; Chapter 19.
5. Daliri, E.B.M.; Lee, B.H.; Park, M.H.; Kim, J.-H.; Oh, D.-H. Novel angiotensin I-converting enzyme inhibitory peptides from soybean protein isolates fermented by *Pediococcus pentosaceus* SDL1409. *LWT Food Sci. Technol.* **2018**, *93*, 88–93. [CrossRef]
6. Aguilar-Toalá, J.E.; Santiago-López, L.; Peres, C.M.; Peres, C.; Garcia, H.S.; Vallejo-Cordoba, B.; Hernández-Mendoza, A. Assessment of multifunctional activity of bioactive peptides derived from fermented milk by specific *Lactobacillus plantarum* strains. *J. Dairy Sci.* **2017**, *100*, 65–75. [CrossRef]
7. Nongonierma, A.B.; Fitzgerald, R.J. The scientific evidence for the role of milk protein-derived bioactive peptides in humans: A Review. *J. Funct. Foods* **2015**, *17*, 640–656. [CrossRef]
8. Wang, X.; Wang, X.M.; Hao, Y.; Teng, D.; Wang, J.H. Research and development on lactoferrin and its derivatives in China from 2011–2015. *Biochem. Cell Biol.* **2017**, *95*, 162–170. [CrossRef] [PubMed]
9. de Almeida, A.J.P.O.; Rezende, M.S.A.; Dantas, S.H.; Silva, S.L.; de Oliveira, J.C.P.L.; de Azevedo, F.L.A.A.; Alves, R.M.F.R.; de Menezes, G.M.S.; dos Santos, P.F.; Gonçalves, T.A.F.; et al. Unveiling the role of inflammation and oxidative stress on age-related cardiovascular diseases. *Oxid. Med. Cell. Longev.* **2020**, *2020*, 1954398. [CrossRef] [PubMed]
10. Garcia, B.D.; de Barros, M.; Rocha, T.D. Bioactive peptides from beans with the potential to decrease the risk of developing noncommunicable chronic diseases. *Crit. Rev. Food Sci. Nutr.* **2020**. [CrossRef]
11. Hernández-Ledesma, B.; Hsieh, C.-C.; de Lumen, B.O. Antioxidant and anti-inflammatory properties of cancer preventive peptide lunasin in RAW264.7 macrophages. *Biochem. Biophys. Res. Commun.* **2009**, *390*, 803–808. [CrossRef]
12. Rajendran, P.; Chen, Y.F.; Chen, Y.F.; Chung, L.C.; Tamilselvi, S.; Shen, C.Y.; Huang, C.Y. The multifaceted link between inflammation and human diseases. *J. Cell. Phys.* **2018**, *233*, 6458–6471. [CrossRef]
13. Singharaj, B.; Pisarski, K.; Martirosyan, D.M. Managing hypertension: Relevant biomarkers and combating bioactive compounds. *Funct. Foods Health Dis.* **2017**, *7*, 442–461. [CrossRef]
14. Capriotti, A.L.; Cavaliere, C.; Piovesana, S.; Sampery, R.; Laganá, A. Recent trends in the analysis of bioactive peptides in milk and dairy products. *Anal. Bioanal. Chem.* **2016**, *408*, 2677–2685. [CrossRef]
15. Fitzgerald, R.J.; Cermeño, M.; Khalesi, M.; Kleekayai, T.; Amigo-Benavent, M. Application of *in silico* approaches for the generation of milk protein-derived bioactive peptides. *J. Funct. Foods* **2020**, *64*, 103636. [CrossRef]
16. Søren Drud, N.; Beverly, R.L.; Qu, Y.; Dallas, D.C. Milk bioactive peptide database: A comprehensive database of milk protein-derived bioactive peptides and novel visualization. *Food Chem.* **2017**, *232*, 673–682.
17. Minkiewicz, P.; Iwaniak, A.; Darewicz, M. BIOPEP-UWM Database of bioactive peptides: Current opportunities. *Int. J. Mol. Sci.* **2019**, *20*, 5978. [CrossRef] [PubMed]
18. Sentandreu, M.A.; Toldrá, F. A rapid, simple and sensitive fluorescence method for the assay of angiotensin-I converting enzyme. *Food Chem.* **2006**, *97*, 546–554. [CrossRef]
19. Quirós, A.; Contreras, M.M.; Ramos, M.; Amigo, L.; Recio, I. Stability to gastrointestinal enzymes and structure–activity relationship of β-casein-peptides with antihypertensive properties. *Peptides* **2009**, *30*, 1848–1853. [CrossRef]
20. Hernández-Ledesma, B.; Dávalos, A.; Bartolomé, B.; Amigo, L. Preparation of antioxidant enzymatic hydrolyzates from α-lactalbumin and β-lactoglobulin. Identification of active peptides by HPLC-MS/MS. *J. Agric. Food Chem.* **2005**, *53*, 588–593. [CrossRef]

21. Hernandez-Ledesma, B.; Miralles, B.; Amigo, L.; Ramos, M.; Recio, I. Identification of antioxidant and ACE-inhibitory peptides in fermented milk. *J. Sci. Food Agric.* **2005**, *85*, 1041–1048. [CrossRef]

22. García-Nebot, M.J.; Recio, I.; Hernández-Ledesma, B. Antioxidant activity and protective effects of peptide lunasin against oxidative stress in intestinal Caco-2 cells. *Food Chem. Toxicol.* **2014**, *65*, 155–161. [CrossRef] [PubMed]

23. Matsufuji, H.; Matsui, T.; Seki, E.; Osajima, K.; Nakashima, M.; Osajima, Y. Angiotensin I-converting enzyme inhibitory peptides in an alkaline proteinase hydrolysate derived from sardine muscle. *Biosci. Biotech. Biochem.* **1994**, *58*, 2244–2245. [CrossRef] [PubMed]

24. Contreras, M.M.; Sánchez-Infantes, D.; Sevilla, M.A.; Recio, I.; Amigo, L. Resistance of casein-derived bioactive peptides to simulated gastrointestinal digestión. *Int. Dairy J.* **2013**, *32*, 71–78. [CrossRef]

25. Liu, L.; Wei, Y.N.; Chang, Q.; Sun, H.J.; Chai, K.G.; Huang, Z.Q.; Zhao, Z.X.; Zhao, Z.X. Ultrafast screening of a novel, moderately hydrophilic Angiotensin-converting-enzyme-inhibitory peptide, RYL, from silkworm pupa using an fe-doped-silkworm-excrement-derived biocarbon: Waste conversion by waste. *J. Agric. Food Chem.* **2017**, *65*, 11202–11211. [CrossRef] [PubMed]

26. Contreras, M.M.; Carrón, R.; Montero, M.J.; Ramos, M.; Recio, I. Novel casein-derived peptides with antihypertensive activity. *Int. Dairy J.* **2009**, *19*, 566–573. [CrossRef]

27. Suetsuna, K.; Ukeda, H.; Ochi, H. Isolation and characterization of free radical scavenging activities peptides derived from casein. *J. Nutr. Biochem.* **2000**, *11*, 128–131. [CrossRef]

28. Kohmura, M.; Nio, N.; Kubo, K.; Minoshima, Y.; Munekata, E.; Ariyoshi, Y. Inhibition of angiotensin-converting enzyme by synthetic peptides of human beta-casein. *Agric. Biol. Chem.* **1989**, *53*, 2107–2114. [CrossRef]

29. Chiba, H.; Yoshikawa, M. Biologically functional peptides from food proteins: New opioid peptides from milk proteins. In *Protein Tailoring for Food and Medical Purposes*; Feeney, R.E., Whitaker, J.R., Eds.; Marcel Dekker: New York, NY, USA, 1986.

30. Mullally, M.M.; Meisel, H.; FitzGerald, R.J. Synthetic peptides corresponding to alpha-lactalbumin and beta-lactoglobulin sequences with angiotensin-I-converting enzyme inhibitory activity. *Biol. Chem. Hoppe-Seyler* **1996**, *377*, 259–261.

31. Martínez-Maqueda, D.; Miralles, B.; Pascual-Teresa, S.; Reverón, I.; Muñoz, R.; Recio, I. Food-derived peptides stimulate mucin secretion and gene expression in intestinal cells. *J. Agric. Food Chem.* **2012**, *60*, 8600–8605. [CrossRef]

32. Loukas, S.; Varoucha, D.; Zioudrou, C.; Streaty, R.A.; Werner, A.; Klee, W.A. Opioid activities and structures of alpha-casein-derived exorphins. *Biochemistry* **1983**, *22*, 4567–4573. [CrossRef]

33. Hatzoglou, A.; Bakogeorgou, E.; Hatzoglou, C.; Martin, P.-M.; Castanas, E. Antiproliferative and receptor binding properties of α- and β-casomorphins in the T47D human breast cancer cell line. *Eur. J. Pharmacol.* **1996**, *310*, 217–223. [CrossRef]

34. Xu, R.J. Bioactive peptides in milk and their biological and health implications. *Food Rev. Int.* **1998**, *14*, 1–16. [CrossRef]

35. Yamamoto, N.; Akino, A.; Takano, T. Antihypertensive effects of peptides derived from casein by an extracellular proteinase from *Lactobacillus helveticus* CP790. *J. Dairy Sci.* **1994**, *77*, 917–922. [CrossRef]

36. Martínez-Maqueda, D.; Miralles, B.; Cruz-Huerta, E.; Recio, I. Casein hydrolysate and derived peptides stimulate mucin secretion and gene expression in human intestinal cells. *Int. Dairy J.* **2013**, *32*, 13–19. [CrossRef]

37. Fernández-Tomé, S.; Martínez-Maqueda, D.; Girón, R.; Goicochea, C.; Miralles, B.; Recio, I. Novel peptides derived from α_{s1}-casein with opioid activity and mucin stimulatory effect on HT29-MTX cells. *J. Funct. Foods* **2016**, *25*, 466–476. [CrossRef]

38. Hernández-Ledesma, B.; Amigo, L.; Ramos, M.; Recio, I. Release of angiotensin converting enzyme-inhibitory peptides by simulated gastrointestinal digestion of infant formulas. *Int. Dairy J.* **2004**, *14*, 889–898. [CrossRef]

39. Dubynin, V.A.; Asmakova, L.S.; Sokhanenkova, N.Y.; Bespalova, Z.D.; Nezavibat'ko, V.N.; Kamenskii, A.A. Comparative analysis of neurotropic activity of exorphines, derivatives of dietary proteins. *Bull. Exp. Biol. Med.* **1998**, *125*, 131–134. [CrossRef]

40. Haq, M.R.U.; Kapila, R.; Saliganti, R. Consumption of beta-casomorphins-7/5 induce inflammatory immune response in mice gut through Th-2 pathway. *J. Funct. Foods* **2014**, *8*, 150–160. [CrossRef]

41. Kayser, H.; Meisel, H. Stimulation of human peripheral blood lymphocytes by bioactive peptides derived from bovine milk proteins. *FEBS Lett.* **1996**, *383*, 18–20. [CrossRef]

42. Brantl, V.; Teschemacher, H.; Bläsig, J.; Henschen, A.; Lottspeich, F. Opioid activities of β-casomorphins. *Life Sci.* **1981**, *28*, 1903–1909. [CrossRef]

43. Trompette, A.; Claustre, J.; Caillon, F.; Jourdan, G.; Chayvialle, J.A.; Plaisancie, P. Milk bioactive peptides and β-casomorphins induce mucus release in rat jejunum. *J. Nutr.* **2003**, *133*, 3499–3503. [CrossRef]

44. Zoghbi, S.; Trompette, A.; Claustre, J.; Homsi, M.E.; Garzón, J.; Jourdan, G.; Scoazec, J.-I.; Plaisancie, P. β-Casomorphin-7 regulates the secretion and expression of gastrointestinal mucins through a μ-opioid pathway. *Am. J. Physiol. Gastrointest. Liver Physiol.* **2006**, *290*, G1105–G1113. [CrossRef] [PubMed]

45. Claustre, J.; Toumi, F.; Trompette, A.; Jourdan, G.; Guignard, H.; Chayvialle, J.A.; Plaisancie, P. Effects of peptides derived from dietary proteins on mucus secretion in rat jejunum. *Am. J. Physiol. Gastrointest. Liver Physiol.* **2002**, *283*, G521–G528. [CrossRef] [PubMed]

46. Yin, H.; Miao, J.; Ma, C.; Sun, G.; Zhang, Y. β-Casomorphin-7 cause decreasing in oxidative stress and inhibiting NF-κB-iNOS-NO signal pathway in pancreas of diabetes rats. *J. Food Sci.* **2012**, *77*, C278–C282. [CrossRef] [PubMed]

47. Osborne, S.; Chen, W.; Addepalli, R.; Colgrave, M.; Singh, T.; Tranc, C.; Day, L. *In vitro* transport and satiety of a beta-lactoglobulin dipeptide and beta-casomorphin-7 and its metabolites. *Food Funct.* **2014**, *5*, 2706–2718. [CrossRef]

48. Saito, T.; Nakamura, T.; Kitazawa, H.; Kawai, Y.; Itoh, T. Isolation and structural analysis of antihypertensive peptides that exist naturally in Gouda cheese. *J. Dairy Sci.* **2000**, *83*, 1434–1440. [CrossRef]

49. Uenishi, H.; Kabuki, T.; Seto, Y.; Serizawa, A.; Nakajima, H. Isolation and identification of casein-derived dipeptidyl-peptidase 4 (DPP-4)-inhibitory peptide LPQNIPPL from gouda-type cheese and its effect on plasma glucose in rats. *Int. Dairy J.* **2012**, *22*, 24–30. [CrossRef]

50. Martínez-Maqueda, D.; Miralles, B.; Ramos, M.; Recio, I. Effect of β-lactoglobulin hydrolysate and β-lactorphin on intestinal mucin secretion and gene expression in human goblet cells. *Food Res. Int.* **2013**, *54*, 1287–1291. [CrossRef]

51. Nongonierma, A.B.; FitzGerald, R.J. Structure activity relationship modelling of milk protein-derived peptides with dipeptidyl peptidase IV (DPP-IV) inhibitory activity activity. *Peptides* **2016**, *79*, 1–7. [CrossRef]

52. Jinsmaa, Y.; Yoshikawa, M. Enzymatic release of neocasomorphin and β-casomorphin from bovine β-casein. *Peptides* **1999**, *20*, 957–962. [CrossRef]

53. Plaisancié, P.; Boutrou, R.; Estienne, M.; Henry, G.; Jardin, J.; Paquet, A.; Léonil, J. β-casein (94-123)-derived peptides differently modulate production of mucins in intestinal goblet cells. *J. Dairy Res.* **2015**, *82*, 36–46. [CrossRef]

54. Tian, M.; Fang, B.; Jiang, L.; Guo, H.; Cui, J.Y.; Ren, F. Structure-activity relationship of a series of antioxidant tripeptides derived from β-lactoglobulin using QSAR modeling. *Dairy Sci. Technol.* **2015**, *95*, 451–463. [CrossRef]

55. Pihlanto-Leppala, A. Bioactive peptides derived from bovine whey proteins: Opioid and ace-inhibitory peptides. *Trends Food Sci. Technol.* **2000**, *11*, 347–356. [CrossRef]

56. Jacquot, A.; Gauthier, S.F.; Drouin, R.; Boutin, Y. Proliferative effects of synthetic peptides from β-lactoglobulin and α-lactalbumin on murine splenocytes. *Int. Dairy J.* **2010**, *20*, 514–521. [CrossRef]

57. Hernández-Ledesma, B.; Recio, I.; Ramos, M.; Amigo, L. Preparation of ovine and caprine beta-lactoglobulin hydrolysates with ACE-inhibitory activity Identification of active peptides from caprine beta-lactoglobulin hydrolysed with thermolysin. *Int. Dairy J.* **2002**, *12*, 805–812. [CrossRef]

58. Brantl, V.; Gramsch, C.; Lottspeich, F.; Mertz, R.; Jaeger, K.-H.; Herz, A. Novel opioid peptides derived from hemoglobin: *Hemorphins. Eur. J. Pharmacol.* **1986**, *125*, 309–310. [CrossRef]

59. Nakamura, Y.; Yamamoto, N.; Sakai, K.; Okubo, A.; Yamazaki, S.; Takano, T. Purification and characterization of Angiotensin I-converting enzyme inhibitors from sour milk. *J. Dairy Sci.* **1995**, *78*, 777–783. [CrossRef]

60. Cheung, H.-S.; Wang, F.-L.; Ondetti, M.A.; Sabo, E.F.; Cushman, D.W. Binding of peptide substrates and inhibitors of angiotensin-converting enzyme. Importance of the COOH-terminal dipeptide sequence. *J. Biol. Chem.* **1980**, *255*, 401–407.

61. Samaranayaka, A.G.P.; Li-Chan, E.C.Y. Food-derived peptidic antioxidant: A review of their production, assessment, and potential applications. *J. Funct. Foods* **2011**, *3*, 229–254. [CrossRef]

62. Circu, M.L.; Aw, T.Y. Reactive oxygen species, cellular redox systems, and apoptosis. *Free Rad. Biol. Med.* **2010**, *48*, 749–762. [CrossRef]
63. Manna, C.; Migliardi, V.; Sannino, F.; De Martino, A.; Capasso, R. Protective effects of synthetic hydroxytyrosol acetyl derivatives against oxidative stress in human cells. *J. Agric. Food Chem.* **2005**, *53*, 9602–9607. [CrossRef]
64. Kim, H.J.; Lee, E.K.; Parke, M.H.; Ha, Y.M.; Juang, K.J.; Kim, M.-S.; Kim, M.K.; Yu, B.P.; Chung, H.Y. Ferulate protects the epithelial barrier by maintaining tight junction protein expression and preventing apoptosis in tert-butyl hydroperoxide-induced Caco-2 cells. *Phytother. Res.* **2013**, *27*, 362–367. [CrossRef] [PubMed]
65. Bautista-Expósito, S.; Peñas, E.; Frias, J.; Martínez-Villaluenga, C. Pilot-scale produced fermented lentil protects against t-BHP-triggered oxidative stress by activation of Nrf2 dependent on SAPK/JNK phosphorylation. *Food Chem.* **2019**, *274*, 750–759. [CrossRef] [PubMed]
66. Indiano-Romacho, P.; Fernández-Tomé, S.; Amigo, L.; Hernández-Ledesma, B. Multifunctionality of lunasin and peptides released during its simulated gastrointestinal digestion. *Food Res. Int.* **2019**, *125*, 108513. [CrossRef] [PubMed]
67. Wang, H.; Joseph, J.A. Quantifying cellular oxidative stress by dichlorofluorescein assay using microplate reader. *Free Rad. Biol. Med.* **1999**, *27*, 612–616. [CrossRef]
68. LeBel, C.P.; Ishiropoulos, H.; Bondy, S.C. Evaluation of the probe 2′,7′-dichlorofluorescin as an indicator of reactive oxygen species formation and oxidative stress. *Chem. Res. Toxicol.* **1992**, *5*, 227–231. [CrossRef] [PubMed]

 *foods*

Article

In Vitro Characterisation of the Antioxidative Properties of Whey Protein Hydrolysates Generated under pH- and Non pH-Controlled Conditions

Thanyaporn Kleekayai [1], Aurélien V. Le Gouic [1], Barbara Deracinois [2], Benoit Cudennec [2] and Richard J. FitzGerald [1,*]

[1] Department of Biological Sciences, University of Limerick, V94 T9PX Limerick, Ireland; thanyaporn.kleekayai@ul.ie (T.K.); aurelien.legouic@ul.ie (A.V.L.G.)

[2] Joint Research Unit BioEcoAgro N° 1158, Université Lille, INRAE, Université Liège, UPJV, YNCREA, Université d'Artois, Université Littoral Côte d'Opale, ICV Institut Charles Viollette, F-59000 Lille, France; barbara.deracinois@univ-lille.fr (B.D.); benoit.cudennec@univ-lille.fr (B.C.)

* Correspondence: dick.fitzgerald@ul.ie; Tel.: +353-61-202-598

Received: 8 April 2020; Accepted: 27 April 2020; Published: 5 May 2020

Abstract: Bovine whey protein concentrate (WPC) was hydrolysed under pH-stat (ST) and non pH-controlled (free-fall, FF) conditions using Debitrase (DBT) and FlavorPro Whey (FPW). The resultant whey protein hydrolysates (WPHs) were assessed for the impact of hydrolysis conditions on the physicochemical and the in vitro antioxidant and intracellular reactive oxygen species (ROS) generation in oxidatively stressed HepG2 cells. Enzyme and hydrolysis condition dependent differences in the physicochemical properties of the hydrolysates were observed, however, the extent of hydrolysis was similar under ST and FF conditions. Significantly higher ($p < 0.05$) in vitro and cellular antioxidant activities were observed for the DBT compared to the FPW–WPHs. The WPHs generated under ST conditions displayed significantly higher ($p < 0.05$) oxygen radical absorbance capacity (ORAC) and Trolox equivalent antioxidant capacity (TEAC) values compared to the FF-WPHs. The impact of hydrolysis conditions was more pronounced in the in vitro compared to the cellular antioxidant assay. WPH peptide profiles (LC-MS/MS) were also enzyme and hydrolysis conditions dependent as illustrated in the case of β-lactoglobulin. Therefore, variation in the profiles of the peptides released may explain the observed differences in the antioxidant activity. Targeted generation of antioxidant hydrolysates needs to consider the hydrolysis conditions and the antioxidant assessment method employed.

Keywords: whey protein hydrolysate; hydrolysis condition; food antioxidant; ORAC; cellular ROS; HepG2; peptides

---

1. Introduction

Whey is a source of bioactive peptides (BAPs) with a range of biological properties including antihypertensive, antimicrobial, antidiabetic as well as antioxidant activities [1,2]. Consumption of whey protein has been linked with beneficial effects on human health, particularly in the prevention and management of metabolic syndrome conditions such as cardiovascular disease, type II diabetes mellitus, obesity and hypertension [3–5]. High intracellular levels of reactive oxygen species (ROS) have been associated with the deleterious modification of cells, nucleic acids (DNA and RNA), proteins and lipids and have also been implicated in accelerating cellular ageing [6]. Cells have different mechanisms to protect themselves from oxidative damage via generation of antioxidant compounds/enzymes, e.g., glutathione, superoxide dismutase (SOD), catalase (CAT) and peroxidase, as well as via the uptake of dietary antioxidants or their precursors [7]. Dietary antioxidants have certain advantages over synthetic antioxidants due to their low risk of side-effects and the fact that they can be included as part of the daily dietary intake [8]. Numerous studies show that whey proteins and their hydrolysates have potential antioxidant effects [1,3–5,9–17]. Therefore, whey proteins may have potential applications as a source of antioxidant activity in the prevention and management of diseases associated with oxidative stress.

Enzymatic hydrolysis is one of the most effective approaches for liberation of BAPs from intact protein sequences [18]. Due to its mild operating conditions, enzyme-catalysed hydrolysis is extensively used for the generation of food-grade protein hydrolysates. The antioxidant properties of whey protein hydrolysates as well as whey-derived BAPs have been reported to display numerous functions including free radical scavenging, hydrogen and electron donation, metal ion chelation, quenching of singlet oxygen, peroxide decomposition and inhibition of lipid oxidation [3,9,10,12,14–17,19]. It is well documented that the hydrolysis conditions, e.g., temperature, pH and ionic strength and type of salt influence the characteristics of the hydrolysates obtained [12,20–26]. The pH of the reaction is considered to be one of the most important parameters during enzymatic hydrolysis. Changes in pH alter the structure of the enzyme as well as its substrate, and consequently can affect enzyme specificity [22,24]. Enzyme specificity determines the resultant peptide profile [24] and, thus, hydrolysate properties [12,20]. The pH can be controlled throughout an hydrolysis reaction by adding acid or base in order to maintain the enzyme at optimum operating conditions. However, this strategy may not be feasible during industrial-scale production. Initially adjusting the pH to the enzyme's optimum value and then allowing the reaction to proceed uncontrolled is often more feasible during the large-scale production of protein hydrolysates.

Le Maux et al. [12] demonstrated the impact of hydrolysis under pH- and non pH-controlled conditions on the physicochemical and bioactive properties of whey protein concentrate hydrolysates (WPHs) generated with papain and papain-like proteases. It was shown that the resultant hydrolysates had a similar degree of hydrolysis (DH) but different peptide profiles. This, in turn, led to differences in hydrolysate bioactive properties. For instance, the hydrolysates obtained under pH-controlled conditions had higher oxygen radical absorbance capacity (ORAC) values compared to the non pH-controlled hydrolysis reaction. Similar trends were subsequently reported by Carvalho et al. [20], where a similar DH but different peptide profiles and surface hydrophobicities were observed following whey protein isolate (WPI) hydrolysis using different hydrolysis conditions.

Generally, protein hydrolysates contain a complex mixture of peptides and amino acids. Therefore, in order to distinguish the antioxidant mechanism(s) of BAPs, it is necessary to employ different antioxidant assays for evaluation of antioxidant potency. Conventionally, assays which measure hydrogen atom transfer (HAT) or electron transfer (ET) are employed in the in vitro assessment of antioxidant activity [27]. For instance, the Trolox equivalent antioxidant capacity (TEAC) assay utilises 2,2′-azino-bis-(3-ethylbenzothiazoline)-6-sulfonic acid (ABTS) and 2,2-diphenyl-1-picrylhydrazyl (DPPH) radicals to measure the HAT and ET activity of test compounds. On the other hand, the ferric reducing antioxidant power (FRAP) is an ET-based assay while the ORAC assay measures the scavenging capacity of test compounds against peroxyl radicals (ROO•). The ORAC assay is considered suitable for assessment of the antioxidant activity of protein hydrolysates as it employs biologically relevant radicals [5]. Furthermore, in situ cell-based assays have been recommended as an approach to evaluate antioxidant activity [1]. Cell-based antioxidant assays include assessment of glutathione peroxidase (GPx), CAT and SOD activities, and oxidative damage of DNA, antioxidant gene expression, inhibition of cellular lipid oxidation, protective effects against oxidatively stressed cells and inhibition of cellular ROS generation [3].

The objective of the present study was to investigate the impact of pH- and non pH-controlled enzymatic hydrolysis conditions on the antioxidant properties of WPHs generated using two enzyme preparations. The antioxidant properties of the WPHs were assessed using the in vitro ORAC assay as well as in situ using oxidatively stressed hepatocyte (HepG2) cell lines. In addition, liquid chromatography coupled with mass spectrometry (LC-MS/MS) was employed to identify some of the WPH peptides potentially responsible for the observed antioxidant activity.

2. Materials and Methods

2.1. Materials

Whey protein concentrate (WPC80, 80.98% ± 0.68% (*w/w*) protein (determined by the Kjeldahl nitrogen determination method)) was obtained from Carbery Group (Balineen, Cork, Ireland). FlavorPro®Whey 750P (>55 casein protease U/g) was obtained from Biocatalysts Ltd. (Cefn, Wales, UK) and Debitrase® HYW20 (11,470 U/g) was obtained from DuPont-Danisco (Marlborough, Wiltshire, UK). ABTS, trifluoroacetic acid (TFA), Trolox, 2,2'-azobis (2-amidinopropane) dihydrochloride (AAPH), mass spectrometry (MS) grade water and acetonitrile were purchased from Sigma-Aldrich (Dublin, Ireland). Sodium hydroxide (NaOH) and high pressure liquid chromatography (HPLC) grade water and acetonitrile were provided by Fisher Scientific (Dublin, Ireland). 2,4,6-Trinitrobenzenesulfonic acid (TNBS) was obtained from Pierce Biotechnology (Medical Supply, Dublin, Ireland). Dulbecco's minimum essential medium (DMEM), fetal bovine serum (FBS), antibiotic-antimycotic solution, L-glutamine, Dulbecco's phosphate buffered saline (PBS), Hank's balanced salt solution (HBSS), 2',7'-dichlorofluorescein-diacetate (DCFH-DA) and dimethyl sulfoxide (DMSO) were purchased from Sigma-Aldrich (Wicklow, Ireland).

2.2. Enzymatic Hydrolysis of WPC80

A 10% (*w/v*) protein solution of WPC80 was prepared by reconstitution in distilled water. The protein suspension was mixed at room temperature (22 ± 2 °C) for 2 h and was then allowed to hydrate overnight (16 h) at 4 °C with gentle agitation. The following day, the protein solution was equilibrated at 50 °C followed by adjustment to pH 7.0 using 2 M NaOH. Hydrolysis was initiated by addition of enzyme at an enzyme to substrate ratio (E:S) of 1.0 and 0.5% (*w/w*) for FlavorPro®Whey 750P and Debitrase® HYW20, respectively. Hydrolysis was carried out at 50 °C for 4 h under gentle agitation. The hydrolysis reaction carried out under ST conditions (Titrando 843, Tiamo 1.4 Metrohm, Dublin, Ireland) was maintained at pH 7.0 for both enzyme preparations. For non pH-controlled conditions (FF), the pH of the solution was monitored throughout the hydrolysis reaction. Aliquots of the hydrolysates were collected at hourly intervals. The reaction was terminated by heating at 80 °C for 20 min. The hydrolysates were then freeze-dried (FreeZone 18 L, Labconco, Kansas City, MO, USA) and stored at −20 °C prior to further analysis.

2.3. Determination of Degree of Hydrolysis (DH)

The extent of hydrolysis was determined in triplicate using the TNBS method, as previously described by Le Maux, et al. [12]. The whey protein hydrolysates (WPHs) and unhydrolysed WPC were diluted with 1% (*w/v*) SDS to obtain 0.1% (*w/v*) protein/protein equivalent solutions. An aliquot (10 μL) was pre-incubated at 50 °C for 30 min prior to mixing with 160 μL of 0.05% (*w/v*) TNBS solution in 0.2125 M sodium phosphate buffer pH 8.2. The absorbance at 350 nm was measured after 1 h of incubation at 50 °C using a microplate reader (BioTek Synergy HT, Waltham, MA, USA). Leucine at different concentrations was used as a standard in order to determine the primary amino group content in the samples. The DH was determined using the following formula:

$$\text{DH } (\%) = \frac{\text{AN}_{\text{WPH}} - \text{AN}_{\text{WPC}}}{\text{Npb}} \times 100 \tag{1}$$

where AN_{WPH} is the amino nitrogen content of the hydrolysate (mg nitrogen/mg protein); AN_{WPC} is the amino nitrogen content of the unhydrolyzed WPC and Npb is the nitrogen content of the peptide bonds in whey protein (123.3 mg/g) [28].

2.4. Sodium Dodecyl Sulfate Polyacrylamide Gel Electrophoresis (SDS-PAGE)

SDS-PAGE of the WPH samples was carried out using Mini-PROTEAN® TGX™ Precast Gels with a polyacrylamide gradient of 4–20% (Bio-Rad Laboratories Inc., CA, USA) under reducing conditions as described by O'Loughlin et al. [29]. The denatured samples, containing 25 µg protein/protein equivalent, along with the broad range (6.5–200 kDa) molecular mass standard (Bio-Rad) were separated using a mini Protean II electrophoresis system (Bio-Rad) at 150 V for 1 h.

2.5. Liquid Chromatography (LC)

The molecular mass distribution and peptide profiles of the freeze-dried WPH samples were analysed using gel permeation high-performance liquid chromatography (GP-HPLC; Waters, Milford, MA, USA) and reverse-phase ultra-performance liquid chromatography (RP-UPLC, Waters), respectively, as previously described by Spellman, et al. [30]. The detector response was monitored at 214 nm. A calibration curve was prepared from the mean retention times of standard proteins and peptides for analysis of molecular mass distribution profiles from the GP-HPLC chromatograms.

2.6. In Vitro Antioxidant Analysis

2.6.1. ORAC Assay

The ORAC assay was performed as described by Le Maux et al. [31]. The WPHs were tested at a final concentration of 0.04 mg/mL. Trolox was used as a positive control at final concentrations ranging from 0.0 to 8.0 µM. The ORAC values were expressed as µmol of Trolox equivalents (TE) per g of freeze-dried sample (FDP, $n = 3$).

2.6.2. TEAC Assay

The TEAC assay measures scavenging activity of the test sample against the ABTS cation radical ($ABTS^{\bullet+}$) as described by Re et al. [32], with some modifications. Samples (10 µL) at a final concentration of 0.04 mg FDP/mL were mixed with the $ABTS^{\bullet+}$ working solution (200 µL) in a 96-well microplate. The $ABTS^{\bullet+}$ was monitored at 734 nm following incubation at 30 °C for 6 min. Trolox was used as a positive control at final concentrations ranging from 0.0 to 50.0 µM. The scavenging activity was reported as µmol TE per g FDP ($n = 3$).

2.7. Cellular Antioxidant Assay

2.7.1. Tissue Culture

HepG2 (ECACC 85011430) cells were maintained in DMEM supplemented with 10% (*v/v*) heat inactivated FBS, 1% (*v/v*) non-essential amino acids, 1% (*v/v*) antibiotic-antimycotic solution (100 U/mL penicillin, 100 µg/mL streptomycin, 0.25 µg/mL amphotericin B) and 2 mM L-glutamine. Cells were incubated at 37 °C in a humidified environment with 5% CO_2. HepG2 cells, at passage number 100–110, were used for the experiments. Cell culture medium were replaced every two days, and cells were sub-cultured at 2–4 day intervals before reaching 85%–90% confluence.

2.7.2. Cell Viability

The HepG2 cells were seeded at 6.0×10^4 cells in 200 µL per well on black 96-well plates (Corning, NY, USA) supplemented with DMEM and incubated for 24 h at 37 °C. The medium (200 µL) was aspirated and the adherent cells were rinsed with PBS. The cells were then treated with the WPHs at final concentrations ranging from 0–12.5 mg/mL, prepared in HBSS, and were incubated at 37 °C for 1 h. Following incubation, the medium containing the test compounds was removed and rinsed with PBS. Cell viability was evaluated by exposure to 10% (*v/v*) PrestoBlue® (Invitrogen, Biosciences, Dublin, Ireland) in DMEM. The fluorescence of reduced-PrestoBlue due to metabolically active cells was then measured at excitation and emission wavelengths of 560 and 635 nm, respectively, using a microplate reader (Biotek) every 10 min for 2 h at 37 °C. Control cells without treatment with WPH samples were also exposed to PrestoBlue®. The analysis was performed in triplicate ($n = 3$) and the results were reported as the percentage of viable cells in the population treated with different WPH concentrations compared to control cells without treatment.

2.7.3. Assay of Intracellular ROS Generation

The cellular antioxidant assay determined the formation of ROS using the oxidation sensitive dye, DCFH-DA, according to the method of Yarnpakdee et al. [33] with some modifications. DCFH-DA was initially prepared at 4 mM in DMSO and was then diluted to 100 µM in HBSS immediately prior to use. The HepG2 cells were seeded at a density of 6.0×10^4 cells in 200 µL per well in black 96-well plates and were then incubated at 37 °C in 5% CO_2 for 24 h. The medium (200 µL) was aspirated and the adherent cells were rinsed with HBSS. DCFH-DA (100 µL) was added to the cells and the plates were incubated at 37 °C, 5% CO_2 for 30 min. The cells were treated with the test samples (100 µL) at concentrations ranging from 0 to 10 mg/mL (final concentration) and incubated for 1 h. A positive control containing Trolox at final concentrations of 50 and 100 µM instead of WPH was carried out under the same conditions. An aliquot (100 µL) of medium containing test compounds was removed and 100 µL of 800 µM AAPH in HBSS was added. The fluorescence (excitation: 485 nm, emission: 535 nm) of the 2′,7′-dichlorofluorescein (DCF) product resulting from the oxidation of DCFH in the presence of ROS was measured using a plate reader (Biotek) every 10 min for 90 min at 37 °C. Negative control wells consisted of cells in the presence of DCFH-DA and AAPH without hydrolysates. The intracellular ROS level obtained in the presence of the WPH test samples was expressed as a percentage of the relative fluorescence intensity of the negative control cells.

2.8. Peptide Identification by LC-MS/MS

Peptide identification was performed in the 4 h hydrolysates using LC-MS/MS as described by Nongonierma et al. [34]. This consisted of an UltiMate 3000 ultra-HPLC (UHPLC) system (Dionex, Camberley, Surrey, UK) coupled with a quadrupole time-of-flight mass spectrometer (Q-TOF, Impact HD™, Bruker Daltonics GmbH, Bremen, Germany) fitted with an electro-spray ionisation (ESI) source operated in positive ion mode. UHPLC peptide separation was performed using an Aeris Peptide XB-C18 column (150 × 2.1 mm, 1.7 µm; Phenomenex, Cheshire, UK) fitted with a security guard UHPLC C18-PEPTIDE. Mass spectra were scanned at acquisition ranges between 50–600 and 100–2500 m/z for short and long peptides, respectively. Data acquisition was performed using Hystar software (Bruker Daltonics) [35].

Peptide identification was performed using PEAKS Studio (version 7.5, Bioinformatics Solutions Inc., Waterloo, Canada) software and its database search tools. The database used was UniProt_SwissProt (http://www.uniprot.org), taxa *Bos taurus*. The false discovery rate (FDR), average local confidence (ALC) and MS/MS tolerance were set at 1%, 90% and 0.3 Da, respectively. The number of unique and common peptides identified in all samples were subsequently presented in Venn diagram format using the InteractiVenn web-based tool [36]. In addition, peptide abundance was visualised in the form of heat map. Briefly, the occurrence of amino acids within the peptides identified specifically originating from β-lactoglobulin (β-lg) were summated. The results were expressed using a colour code where high, low and no occurrence of an individual amino acid were represented in red, yellow and white, respectively.

Statistical analysis of the peptide maps generated from the LC-MS/MS data acquired from the long peptide detection method was performed using Progenesis QI software for proteomics (Version 4.0, Waters, Milford, MA). The data was subjected to successive processing as follows: (i) alignment of the peptide maps, (ii) peak picking with an intensity threshold set at 2000 and a maximum charge set at 6 and (iii) data standardisation in order to perform statistical analysis on the main components using principle component analysis (PCA). The variables used were derived from the comparison of peptide maps, i.e., the position of the isotopic mass and its intensity.

The peptides identified in the WPHs were searched against the current literature as well as by using the BioPEP-UWM (http://www.uwm.edu.pl/biochemia/index.php/en/biopep) and PepBank (http://pepbank.mgh.harvard.edu/) databases for the presence of previously reported bioactive properties. The location of the identified peptides within the mature bovine β-lg sequence was obtained from Protein BLAST on the National Center for Biotechnology Information (NCBI) resource portal (https://blast.ncbi.nlm.nih.gov/Blast.cgi).

2.9. Statistical Analysis

Statistical analysis was performed using IBM SPSS Statistics 24 (IBM, Chicago, IL, USA). The results were analysed by one-way analysis of variance (ANOVA) or student t-test at a significance level of $p < 0.05$. Where applicable, multiple comparisons were performed using Tukey's post-hoc test.

3. Results and Discussion

3.1. Degree of Hydrolysis (DH) of WPHs

The DH's achieved as a function of incubation time for the WPHs generated using FlavorPro Whey (FPW) and Debitrase (DBT) under pH- and non pH-controlled conditions are shown in Figure 1. DH increased with incubation time for both enzyme preparations. No major impact of the hydrolysis conditions (ST vs. FF) on the extent of hydrolysis was evident. In general, the DBT–WPHs had a higher extent of hydrolysis compared to the FPW–WPHs with DH values of ~14% and 8%, respectively, being reached following 4 h incubation. The pH of the hydrolysate solutions during FF conditions decreased to ~pH 6.7 and 6.2 for the WPHs generated using FPW (Figure 1a) and DBT (Figure 1b). This decrease in pH during FF hydrolysis is due to the release protons (H$^+$) during the cleavage of peptide bonds. Similar DH values for WPHs generated under ST and FF conditions have been reported previously by Le Maux et al. [12] and Carvalho et al. [20].

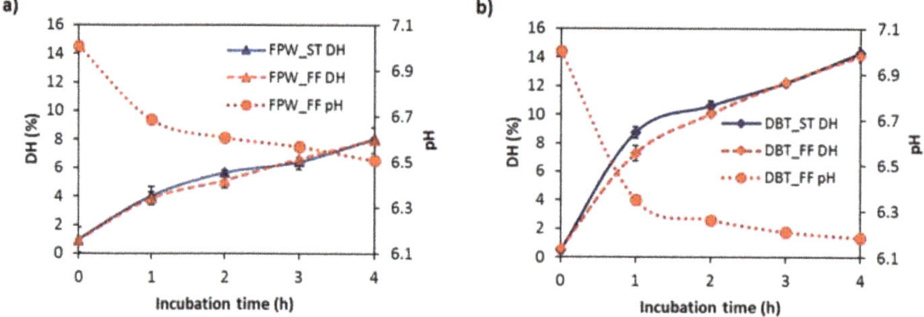

Figure 1. Degree of hydrolysis (DH) and pH profiles as a function of incubation time at 50 °C during the hydrolysis of whey protein concentrate using (**a**) FlavorPro Whey (FPW) and (**b**) Debitrase (DBT) under pH- and non pH-controlled conditions (ST and FF, respectively).

Both enzyme preparations used during WPC hydrolysis contain microbial proteinase activities enriched with exopeptidase activity. FlavorPro Whey 750P derived from *Aspergillus spp.* is reported to cleave at L, F, K, M, E, V, T and C residues (Biocatalyst Technical Bulletin Revision 2: 24 September 2014). Debitrase HYW20 derived from *Aspergillus oryzae* and *Bacillus spp.* possesses leucine aminopeptidase (LAP) and post-proline dipeptidyl aminopeptidase activities [37]. The fact that Debitrase contains proteases from *Bacillus spp.*, which generally have broad specificity [38], may have contributed to the higher extent of hydrolysis observed in the hydrolysates with this enzyme. The application of both enzyme preparations (FPW and DBT) has previously been reported to reduce bitterness in milk protein hydrolysates [39,40].

Analysis of DH only provides an indication of the overall extent of peptide bond cleavage compared to the unhydrolysed sample. However, it does not give any information on the mechanism of hydrolysis or on which peptide bonds were hydrolysed [26]. Therefore, the observation of similar DH values between the ST and FF conditions for each enzyme preparation does not imply similar cleavage sites during WPC hydrolysis. In addition, not all the cleavage sites are hydrolysed at the same rate. This is due to the fact that the rate of hydrolysis of a specific cleavage site is affected by the presence of other amino acids (subsite) in the position adjacent to the cleavage site [24,26,41]. Therefore, further investigation on the impact of hydrolysis conditions on the peptide profile and the antioxidant properties of the WPHs was carried out herein.

3.2. Electrophoresis and Molecular Mass Distribution Profiles

The electrophoretic profiles showed the degradation of protein bands corresponding to the major intact whey proteins (β-lg and α-lactalbumin (α-la)), as well as the generation of low molecular mass compounds <6.5 kDa in the WPHs. This degradation was influenced by the enzyme preparation used for hydrolysis and the hydrolysis conditions, as shown in Figure 2. However, it was noted that a band corresponding to bovine serum albumin (BSA) was observed throughout the incubation period for all samples, albeit with lower intensity compared to that in unhydrolysed WPC. The electrophoretic profiles also show that the WPHs generated using FPW displayed a limited extent of hydrolysis of the main whey protein bands which agrees with the lower extent of WPC hydrolysis compared to that observed in the DBT–WPHs (Figures 1 and 2).

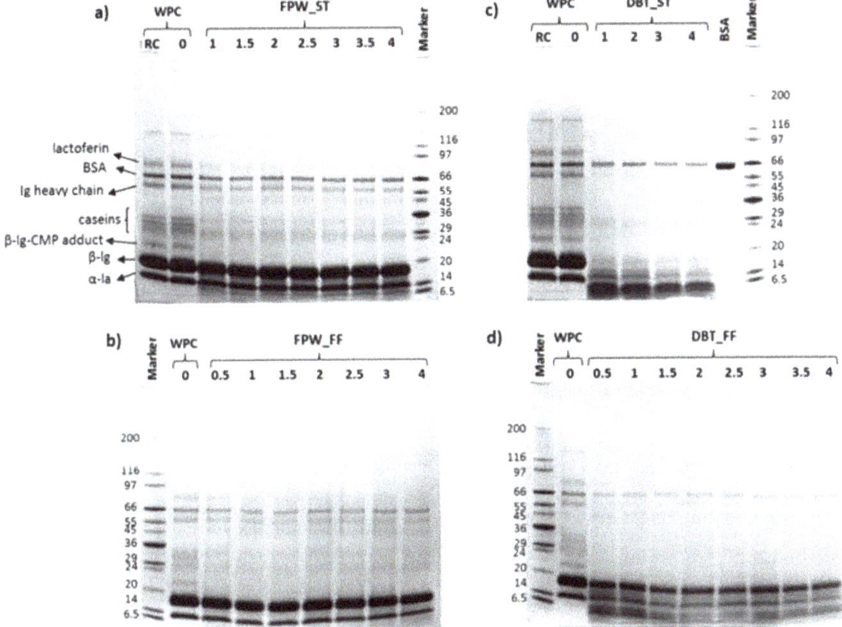

Figure 2. Gel electrophoresis profiles of the reconstituted (RC) intact whey protein concentrate (WPC80), WPC 0 h (WPC0) and the whey protein hydrolysates (WPHs) generated using (**a**) FlavorPro Whey under pH control (FPW_ST), (**b**) FPW without pH control (FPW_FF), (**c**) Debitrase under pH control (DBT_ST) and (**d**) DBT without pH control (DBT_FF) as a function of incubation time (h) at 50 °C. BSA: bovine serum albumin; CMP: caseinomacropeptide; β-lg: β-lactoglobulin; α-la: α-lactalbumin.

In addition, these results highlighted that the WPHs generated under different hydrolysis conditions (ST vs. FF) despite having similar DH values (Figure 1), showed different WPC digestion profiles, particularly in the case of DBT–WPHs (Figure 2c,d). Previous studies by Le Maux et al. [12], Carvalho et al. [20] and Fernández and Kelly [23] also reported different digestion profiles between whey protein hydrolysed under ST and FF conditions (while having comparable DH values). Butré et al. [24] demonstrated significant changes in enzyme selectivity (up to 80%) toward cleavage sites in β-lg as a function of pH which also resulted in different hydrolysate molecular mass distribution profiles. This indicates that the kinetics of peptide release were influenced by the changes in pH during the FF conditions. The changes in enzyme selectivity were previously attributed to modifications in the charge state of amino acids at the active site of the enzyme and at the site of cleavage, as well as in the region adjacent to the cleavage sites [24].

The molecular mass distribution profiles (Figure 3) obtained following GP-HPLC of the WPHs displayed similar results to those observed in the electrophoresis profiles. A greater proportion of high molecular mass components (>10 kDa) was observed in the FF_WPHs compared to the ST_WPHs during hydrolysis with both enzyme preparations. In contrast, Le Maux et al. [12] reported a higher proportion of high molecular mass components (>10 kDa) in ST generated hydrolysates compared to FF conditions for WPC hydrolysates generated with papain. However, in the case of WPHs generated with papain-like activity there was no major differences between the molecular mass profiles of the ST and FF hydrolysates. This indicated that the effect of hydrolysis conditions on the molecular mass distribution profiles was enzyme-dependent. These differences may be explained by the lower optimum pH range of papain, i.e., between pH 5–8, when compared to DBT and FPW which have optimum pH values between pH 6–8.

A general correlation between the molecular mass distribution profiles and DH was evident in that the proportion of peptides <1 kDa increased as a function of incubation time in all the WPH samples. The DBT–WPHs which had higher DHs than the FPW–WPHs had a higher proportion of low (<1 kDa) molecular mass components (Figure 3). The relatively high proportion of high molecular mass components (>10 kDa) in all hydrolysates (ranging between 20% and 70%) may be related to a relatively low level of broad specificity proteinase activities in the FPW and DBT preparations. This may be related to the fact that these enzymes are primarily marketed as exopeptidase containing preparations for protein hydrolysate debittering applications.

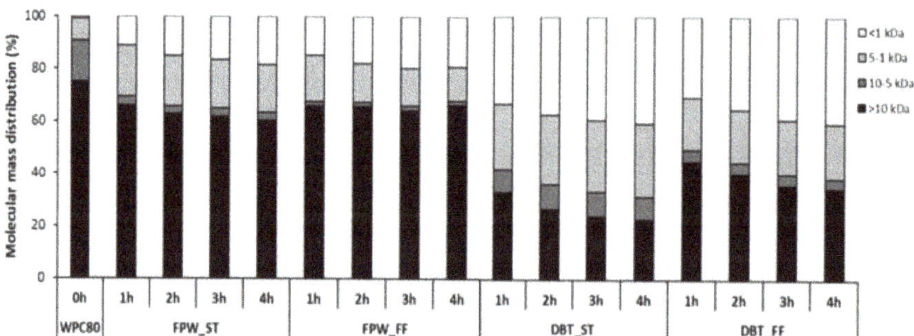

Figure 3. Molecular mass distribution profile of unhydrolysed whey protein concentrate (WPC80) and the whey protein hydrolysates (WPHs) generated using FlavorPro Whey (FPW) and Debitrase (DBT) under pH- and non pH-controlled conditions (ST and FF, respectively) during the course of a 4 h hydrolysis period.

3.3. Reverse-Phase (RP) Peptide Profiles

The peptide profiles of the hydrolysates generated are shown in Figure 4. The degradation of the intact proteins in the WPC as well as the release of hydrophilic peptides was observed with both enzyme preparations, however, this was more pronounced in the case of DBT–WPHs. In addition, the hydrolytic enzymes had a major impact on the peptide profiles of the WPHs with a greater extent of hydrolysis of the intact whey proteins being observed in the DBT–WPHs (Figure 4c,d). Furthermore, the DBT_ST contained a limited amount of intact whey proteins while the DBT_FF hydrolysates had some remaining intact β-lg (Figure 4d). The influence of hydrolysis conditions (ST vs. FF) on peptide profiles concurs with previous reports by Le Maux et al. [12], Butré et al. [24] and Carvalho et al. [20]. Butré et al. [24] showed that, at similar DH, WPI hydrolysed under different constant pH values (pH 7.0–9.0) resulted in different concentrations of residual intact proteins, including β-lg, in the hydrolysates. Le Maux et al. [12] reported that WPC hydrolysed using papain or papain-like activity with ST and FF conditions had comparable overall peptide profiles with different intensities in some peptide peaks. In addition, Fernández and Kelly [23] suggested that different hydrolysis conditions resulted in different kinetics of peptide release, where a slower reaction rate occurred in the FF in comparison to the ST conditions. Carvalho et al. [20] demonstrated that whey protein hydrolysates generated without pH control exhibited significantly higher surface hydrophobicities than those produced under pH control. Therefore, the change in pH during FF hydrolysis may lead to changes in enzyme cleavage specificity resulting in different peptide profiles being observed in comparison to those of the hydrolysates generated under ST.

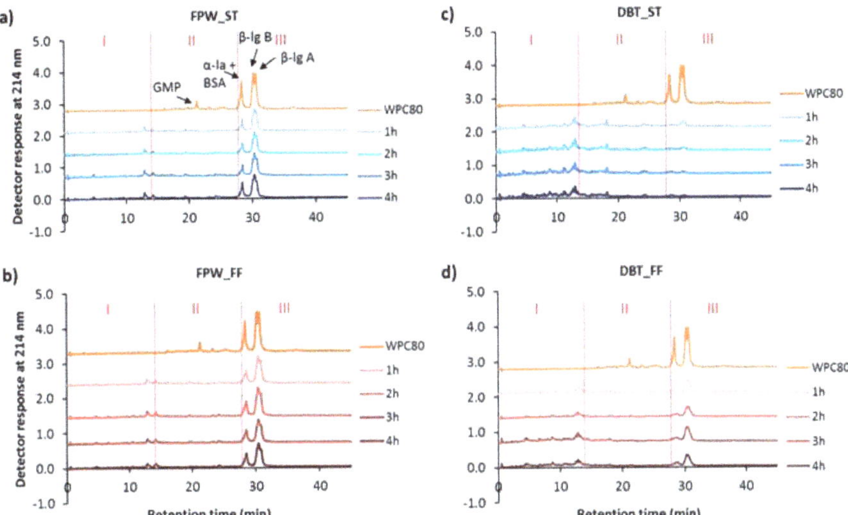

Figure 4. Reverse-phase ultra-performance liquid chromatographic (RP-UPLC) profiles of unhydrolysed whey protein concentrate (WPC80) and the whey protein hydrolysates (WPHs) generated using (**a**) FlavorPro Whey under pH controlled (FPW_ST), (**b**) FlavorPro Whey under non pH controlled (FPW_FF), (**c**) Debitrase under pH controlled (DBT_ST) and (**d**) Debitrase under non pH controlled (DBT_FF) conditions during the course of 4 h hydrolysis. Regions labelled I, II and III represent linear acetonitrile gradients between 0–20%, 20%–40% and >40%, respectively. β-lg: β-lactoglobulin; α-la: α-lactalbumin; BSA: bovine serum albumin; GMP: glycomacropeptide.

3.4. In Vitro Antioxidant Properties

The in vitro antioxidant properties of the WPHs generated with FPW and DBT under ST and FF conditions were assessed using the TEAC and ORAC assays and the results are shown in Figure 5a–d. The TEAC and ORAC values of the WPHs were in the range of 76.0–250.6 and 113.3–403.9 μmol TE/g, respectively, depending on the enzyme and hydrolysis conditions used. In general, the antioxidant properties of all hydrolysates were significantly higher than unhydrolysed WPC with the exception of the TEAC values for the FPW_FF WPHs (Figure 5). These results were generally comparable to the previously reported by other studies. Le Maux et al. [31] reported ORAC values ranging from 179.5–227.6 µmol TE/g for Protamax-WPHs. Power et al. [42] reported an ORAC value for a β-lg tryptic hydrolysate of 467.65 µmol TE/g dry weight.

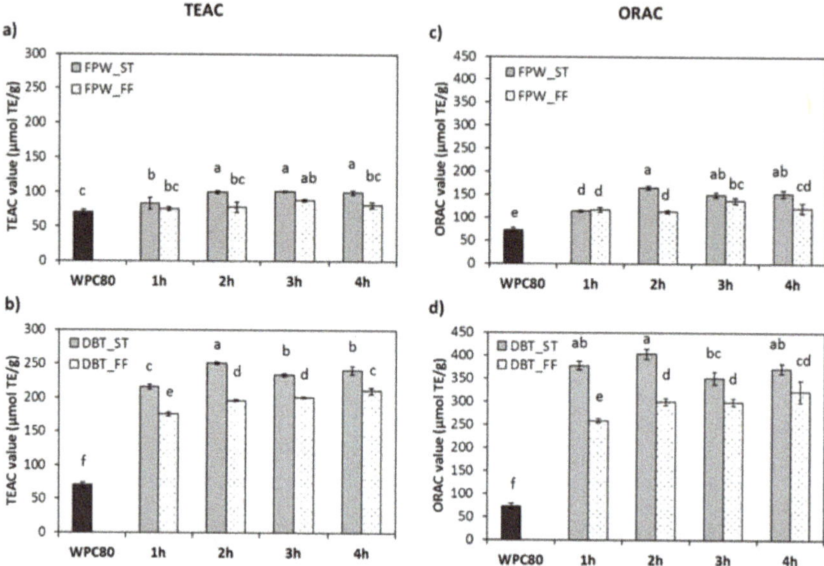

Figure 5. In vitro antioxidant activities reported as Trolox equivalent antioxidant capacity (TEAC) and oxygen radical absorbance capacity (ORAC) of unhydrolysed whey protein concentrate (WPC80) and the whey protein hydrolysates (WPHs) generated using (**a**) and (**c**) FlavorPro Whey under pH- and non pH-controlled (FPW_ST and FPW_FF, respectively) and (**b**) and (**d**) Debitrase under pH- and non pH-controlled (DBT_ST and DBT_FF, respectively) conditions during the course of 4 h hydrolysis period. Values reported are mean ± SD, $n = 3$. Different letters indicated significant difference at $p < 0.05$.

No major further changes in the antioxidant activities of all hydrolysates occurred after 1 h incubation with the enzymes (Figure 5). This may be due to the higher rate of hydrolysis in the first hour of incubation (as shown in Figure 1), for both enzyme preparations. These results were confirmed by the molecular mass distribution and the peptide profiles (Figures 3 and 4). The lower antioxidant activities in the FPW–WPHs may be associated with the lower extent of hydrolysis in these samples. This finding indicates that the WPHs generated by DBT had more efficient hydrogen atom and electron transfer abilities (ABTS assay) as well as scavenging activity against peroxyl radicals (ORAC assay) than the FPW–WPHs. A number of previous studies also demonstrated an enzyme-dependent effect on the in vitro antioxidant potencies of WPHs. Mann et al. [43] reported the antioxidant activity of WPHs generate using Flavourzyme, Alcalase or Corolase PP having TEAC values ranging from 0.81–1.42 µM TE/mg protein. They also suggested that the TEAC values of the WPHs were associated with the extent of hydrolysis which was also in agreement with the results of Salami et al. [17] when hydrolysing whey proteins using Proteinase K, thermolysin, trypsin or chymotrypsin. A significant increase in DPPH scavenging activity as a function of incubation time was observed in the WPH generated using a serine protease from *Myceliophthora thermophile*, while ferric chelation activity did not change after 3 h incubation ($p > 0.05$). No significant change in antioxidant properties was observed in the case of WPH hydrolysed with a metalloprotease from *Eupenicillium javanicum* [14]. These results therefore indicated that the antioxidant properties of WPHs are influenced by the enzyme preparation used which may in turn be linked to the specificity and the extent of hydrolysis achieved. However, O'Keeffe and FitzGerald [15] reported that the ORAC values of a 5 kDa permeate fraction of WPC hydrolysed with Alcalase, Neutrase, Corolase PP or Flavourzyme were not significantly ($p > 0.05$) different (0.6–0.9 µmol TE/mg sample), while the DH ranged between 11.4%–20.5%.

The results showed that the ST conditions resulted in significantly higher antioxidant activities compared to the FF conditions for both enzyme preparations following 4 h incubation (Figure 5). This may be linked to the different physicochemical characteristics, i.e., electrophoretic, molecular mass distribution and reverse-phase profiles, of the ST- vs. FF-WPHs generated using the same enzyme preparation (Figures 2–4). Le Maux et al. [12] reported ORAC values for WPHs ranging from 193–308 µmol TE/g depending on the enzyme and the hydrolysis condition used. The highest ORAC value was found in the WPHs generated under ST (pH 7.0) conditions which was significantly higher than for the WPHs generated at either a lower constant pH (pH < 7.0) or during FF conditions ($p < 0.05$). Therefore, these results showed that controlling the pH at enzyme optimum values can contribute to the release of more potent antioxidant peptides at least when hydrolysing WPC with FPW and DBT.

3.5. Cellular Antioxidant Activity

Biochemical antioxidant assays are considered as generic in vitro assays where the results obtained may not be readily translated to more complex systems such as in the human body [43]. Therefore, it is useful to assess antioxidant properties using in situ cellular-based assays which may be more representative of the target site of oxidative stress in vivo. Several studies reported on the application of cellular antioxidant-based assays of whey protein and its derivatives using various cell lines, e.g., mice myoblast (C2C12) [44,45], human lung fibroblast (MRC-5) [46], rat pheochromocytoma (PC12) [47], human colonic adenocarcinoma (Caco-2) [48], human umbilical vein endothelial (HUVECs) [15] and human tracheobronchial epithelial (1HAEo) [49] as well as human hepatocyte (HepG2) cells [45,50], as reviewed by Corrochano et al. [3].

The cell cytotoxicity of two oxidative stress inducers (AAPH and H_2O_2) was pre-evaluated herein at concentrations ranging from 0–1,000 µM in order to investigate their potential toxic effects on HepG2 cells. The results showed that both inducers resulted in similar effects on cell viability (Supplementary Data, Figure S1). A toxic effect yielding <70% cell viability was found at levels >700 µM for both type of inducer. Due to the similar effects observed between AAPH and H_2O_2, AAPH was selected to represent the oxidative stress inducer at a concentration of 800 µM giving cell viability at a minimum of 50%. In addition, peroxyl radicals from AAPH are reported to have a longer half-life than hydroxyl radicals generated from H_2O_2 at 10^{-2} and 10^{-9} s, respectively [51]. Furthermore, the cell viability using PrestoBlue® of the 4 h-WPHs was evaluated at different concentrations up to 12.5 mg/mL. The WPHs did not appear to have an impact on cell viability with >99% viability (Supplementary Data, Figure S2). Therefore, 3 concentrations of the WPHs, i.e., 1, 5 and 10 mg/mL (final concentrations), were selected for further investigation on their cellular antioxidant activity effects.

The cellular antioxidant assay was carried out to assess the reduction effect of the test samples against AAPH induced intracellular ROS generation, as per Yarnpakdee et al. [33]. The commercial antioxidant, Trolox, was used as a positive control. As expected, Trolox led to a significant reduction in ROS generation compared to the negative control, i.e., AAPH-stressed cells, which was considered to yield 100% ROS generation (Figure 6). Treatment of the cells with 50 or 100 µM Trolox did not show significant differences in the reduction of ROS generation (with 81.09% ± 9.26% and 61.65% ± 4.06%, respectively). Intracellular ROS generation in the WPHs treated cells was in the range 20%–78%. The result therefore showed that treatment with the WPHs led to lower levels of ROS generation. The FPW–WPHs had limited effect on ROS generation (59.2%–78.2%) compared to the DBT–WPHs which showed a greater range in the reduction of ROS generation (19.7%–75.9%).

The DBT_ST WPH at 10 mg/mL exhibited the most potent cellular ROS generation reducing activity giving an ~80% reduction in intracellular ROS generation in AAPH-stressed HepG2 cells in comparison to the control. Honzel et al. [52] associated such a strong inhibition of cellular ROS formation with anti-inflammatory properties. A WPI hydrolysate generated with the aid of high pressure pre-treatment (at 550 MPa) was reported to inhibit the effects of the pro-inflammatory cytokine (IL-8) and ROS generation by up to 50% and 76%, respectively, in H_2O_2-stressed Caco-2 cells in a dose-dependent manner [48]. Likewise, Bamdad et al. [9] reported on high hydrostatic pressure assisted β-lg hydrolysates (BLGHs) displaying an improvement in antioxidant activity and anti-inflammatory properties in lipopolysaccharide (LPS)-stimulated RAW264.7 macrophage cells. The antioxidant activity of the BLGHs was also enzyme-dependent. In the case of anti-inflammatory properties, the BLGHs reduced the nitric oxide level and showed suppression of pro-inflammatory cytokines (tumor necrosis factor (TNF-α) and IL-1β) in LPS-stimulated RAW264.7 cells.

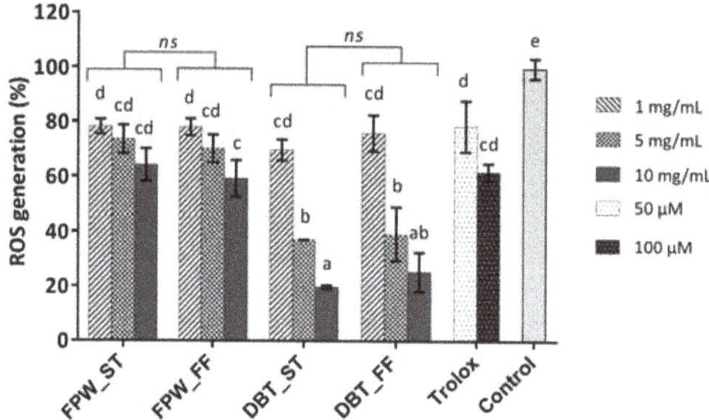

Figure 6. Extent of intracellular reactive oxygen species (ROS) generation in 2,2′-azobis (2-amidinopropane) dihydrochloride (AAPH) stressed-HepG2 cells treated with 1–10 mg/mL (final concentration) of the 4 h whey protein hydrolysates generated using FlavorPro Whey (FPW) and Debitrase (DBT) under pH- (ST) and non pH-controlled (FF) conditions. Trolox at 50 and 100 µM was used as a positive control. Values reported are mean ± SD, $n = 3$. Different letters indicated significant difference at $p < 0.05$. ns: non-significant ($p \geq 0.05$).

With the exception of DBT-FF WPH at 5 and 10 mg/mL, the reduction of intracellular ROS generation of the DBT–WPHs treated cells was observed to occur in a dose-dependent manner ($p < 0.05$). The greater reduction in ROS generation in the HepG2 cells treated with the DBT–WPHs concurred with the results observed in the in vitro TEAC and ORAC assays (Figure 5b and d, respectively). This may be associated with the higher extent of hydrolysis in these samples (Figure 1). The hydrolysis conditions (ST vs. FF) did not have a major impact on intracellular ROS generation for the hydrolysates generated with either DBT or FPW (Figure 6). To our knowledge, this is the first report demonstrating the contribution of hydrolysis conditions on the cellular antioxidant activity of WPHs.

Nonetheless, Honzel et al. [52] indicated that the magnitude of reduction in intracellular ROS generation was not directly correlated with ORAC assay values. In addition, the cell-based antioxidant activity also depends on the permeability of the test sample [53] as well as the interaction between the test sample and complex enzyme reactions in biological systems [52]. Kong, et al. [46] reported that WPHs enhanced SOD, GPx and CAT activities in H_2O_2-induced MRC-5 cells. Similar results were also observed by O'Keeffe and FitzGerald [15], where WPHs obtained following membrane filtration (5 kDa) resulted in an increase in the expression of glutathione and CAT activity in HUVECs. On the other hand, the level of antioxidative biomarkers, i.e., glutathione pyruvate transaminase, alkaline phosphatase and creatinine in HepG2 cells decreased in the presence of WPHs [54]. The 3 kDa permeate fraction of a peptic-digest of whey protein derived from buffalo colostrum restored the level of ROS, H_2O_2 and CAT to normal. In addition, it replenished the glutathione level and moderately restored lysosomal enzyme activity in 2,4-ditrophenol (DNP)-induced oxidatively stressed human blood samples [55].

3.6. Peptide Identification by LC-MS/MS

In order to investigate the hydrolysis pattern and enzyme specificities in the FPW/DBT_ST and FPW/DBT_FF hydrolysates, the digests obtained following 4 h incubation were selected for peptide identification by LC-MS/MS. PCA was performed on the mass spectrometry data, more particularly on the detected ions (Figure 7a). The first two dimensions in the PCA explained 61.37% of the variance (detected ions corresponding to variables). The more distant the groups were, the more different they were in terms of ion population. The PCA clearly showed a different ion population and thus a different peptide population between the FPW and DBT hydrolysates.

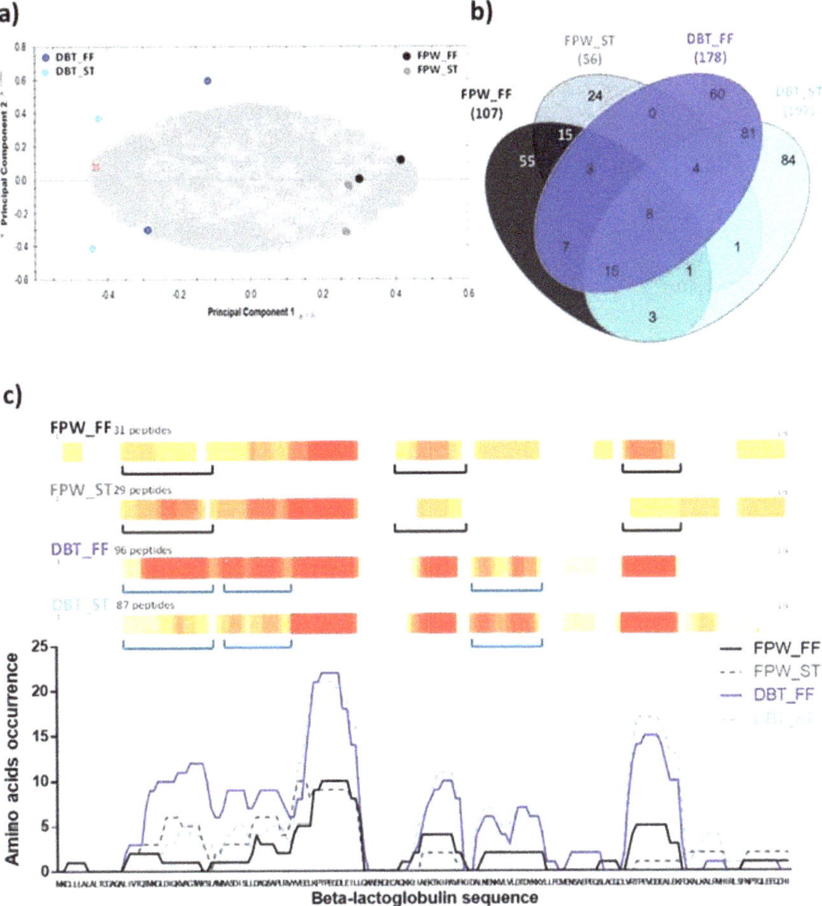

Figure 7. Overall number of common and unique peptides identified in the 4 h whey protein hydrolysates generated using FlavorPro Whey (FPW) and Debitrase (DBT) under pH- and non pH-controlled conditions (ST and FF, respectively). (**a**) Principal component analysis based on comparison of mass spectrometry detected peptides in the WPHs, (**b**) Venn diagram representing the overall number of identified peptides in the WPHs with the common peptide identified in several samples represented by the overlapping area in the diagram and (c) Heat map showing amino acid occurrence in the identified peptides in the primary sequence of β-lactoglobulin where the red colour represents high number of identifications, yellow represents low number of identifications and white represents unidentified amino acids. Underlined regions highlight peptide pattern differences between ST and FF conditions for FPW (in black) or DBT (in blue) hydrolysis.

The total number of peptides identified in the 4 h FPW_FF, FPW_ST, DBT_FF and DBT_ST WPH samples was 107, 56, 178 and 197, respectively. The common and unique peptides in all the samples were identified and the results are presented in Venn diagram format in Figure 7b. This analysis showed that the 4 WPHs, which were derived from two different enzyme preparations and two different hydrolysis conditions, contained eight common peptides. Each hydrolysate sample also had unique peptides, i.e., 55, 24, 60 and 84 peptides in FPW_FF, FPW_ST, DBT_FF and DBT_ST, respectively, (Figure 7b). The higher number of peptides identified in the DBT-WPH samples may be linked to the higher DHs in these samples compared to the FPW–WPHs (Figure 1). However, no clear pattern could be observed in the Venn diagram concerning the effect of hydrolysis conditions on the peptides released. Therefore, the hydrolysis pattern of the major intact whey protein, β-lg, was assessed and the results are presented using a heat map diagram as shown in Figure 7c. This analysis clearly showed that the hydrolysis conditions, i.e., ST vs. FF for the same enzyme preparation, resulted in different cleavage patterns on β-lg (Figure 7c and Table 1). These results are in agreement with those previously reported by Butré et al. [24] on the effect of hydrolysis conditions on peptide profile. Furthermore, the occurrence of specific amino acids in the peptides released was hydrolysis condition-dependent, as illustrated in the regions underlined on the heat map (Figure 7c). This finding may help to explain the differences in the observed antioxidant properties of the hydrolysates herein.

The β-lg-derived peptides identified in the hydrolysates with <10 amino acid residues and containing antioxidant peptide features or having related sequences to those which were previously reported to be bioactive are presented in Table 1. The majority of the peptides identified in the FPW–WPHs were long sequences, i.e., with >56% and >75% of all peptides identified having >10 amino acid residues in the FPW_FF and FPW_ST WPHs, respectively (data not shown). This may be associated with the relatively low extent of hydrolysis in these samples (Figure 1). Some of the peptides identified in the WPHs obtained in the present study have been previously reported to possess antioxidant activity. For instance, LDTDYKK (β-lg f(95–101)) was present in WPC enriched in β-lg when hydrolysed with Corolase PP and thermolysin exerted ORAC antioxidant activity [10]. VLDTDYK (β-lg f(94–100)) and VRTPEVDDE (β-lg f(123–131)), derived from Alcalase hydrolysed cheese whey, had ABTS$^{\bullet+}$ scavenging activity [56]. GTWYSL (β-lg f(17–22)), AMAASDISLL (β-lg f(23–32)), MAASDISL (β-lg f(24–32)) and IIAEKTKIPAVF (β-lg f(71–82)) identified in Alcalase hydrolysed β-lg under high hydrostatic pressure also showed ferric reducing antioxidant activity [9]. In addition, TPEVDDEALEK (β-lg f(125–135)) which was identified in all four hydrolysates in the present study (Table 1) was previously found in WPC hydrolysed with Flavourzyme and Corolase PP exerted ABTS$^{\bullet+}$ scavenging activity [57]. The same peptide was reported in a tryptic β-lg hydrolysate and had ORAC activity (0.004 μmol TE/μmol peptide) [42]. Among the β-lg-derived peptides identified in the present samples, three peptides (VLDTDYK, VRTPEVDDEALEK and TPEVDDEALEK) were not only reported to be resistant to in vitro gastrointestinal digestion, but they also had the capability to be transported across the intestinal epithelium (Caco-2 monolayers) [58].

In general, enzyme preparation plays a key role in peptide release during enzymatic hydrolysis of food proteins and consequently influences hydrolysate bioactive potency. In the present study, two enzyme preparations were used to hydrolyse WPC which led to different profiles of peptides released. For example, LDAQSAPLR (β-lg f(32–40)) and DAQSAPLRVY (β-lg f(33–42)) which were identified in FPW_ST and FF and in DBT_ST and FF, respectively (Table 1), clearly illustrates the differences in the cleavage specificities between the two enzyme preparations. Cleavage post Leu occurred in the case of DBT yielding f(33–42). This may be linked to the presence of LAP in *A. oryzae* [37], while FPW cleaved post Arg in this region of the β-lg molecule (Figure 8).

Table 1. Selected β-lactoglobulin (β-⁻g)-derived peptides identified in the 4 h whey protein hydrolysates (WPHs) generated using FlavorPro Whey (FPW) and Debitrase (DBT) under pH- and non pH-controlled conditions (ST and FF, respectively) and their bioactive properties as previously reported in the literature (including related peptide sequences).

Peptide Sequence [1]	Region in Mature β-lg	Identified In				Reported Activity [2]	IC_{50}/EC_{50} Values [3]	Reference	References to Related Peptide Sequence [1,2,3,4]
		FPW_FF	FPW_ST	DBT_FF	DBT_ST				
KVAGTWYSL	f(14–22)			✓	✓	-	-	-	VAGTWY (ORAC 5.63 µmol TE/µmol peptide, DPP-IVi IC_{50} 74.9 µM [42], ACEi IC_{50} 1,682 µM [59]), VAGT (ACEi IC_{50} 610.3 µM; ORAC 1.66 µmol TE/mmol peptide [60]), WYSL (DPPH & superoxide scavenging activity with EC_{50} 273.63 & 558.42 µM, respectively [61])
VAGTW	f(15–19)				✓	ACEi	534	Pihlanto-LeppÄLÄ, et al. [62]	
GTWYSL	f(17–22)			✓		FRAP	na	Bamdad, et al. [9]	
AMAASDISLL	f(23–32)			✓		FRAP	na	Bamdad, et al. [9]	AASDISLLDAQSAPLR (antibacterial [63]) MAA (ORAC EC_{50} 0.33 µmol TE/mmol peptide, ACEi IC_{50} 515.5 µM [60])
MAASDISLL	f(24–32)			✓		FRAP	na	Bamdad, et al. [9]	
LDAQSAPLR	f(32–40)	✓	✓			ACEi	635	Pihlanto-LeppÄLÄ, et al. [62]	
DAQSAPLRVY	f(33–42)			✓	✓	ACEi	13	Tavares, et al. [64]	
DAQSAPLR	f(33–40)			✓	✓	-	-	-	
SAPLR	f(36–40)			✓		-	-	-	
ELKPTPEGDLEIL	f(45–57)			✓	✓	FRAP	na	Bamdad, et al. [9]	-
LKPTPEGDLEIL	f(46–57)				✓	DPP-IVi	57	Lacroix and Li-Chan [65]	-
IIAEKTKIPAVF	f(71–82)			✓	✓	FRAP	na	Bamdad, et al. [9]	IIAEK (ORAC 0.016 µmol TE/µmol peptide, ACEi IC_{50} 63.7 µM [42]), IAEKTKIP (ORAC [10]), IPAVFK (ACEi IC_{50} 144.8 µM, DPP-IVi IC_{50} 149.1 µM, ORAC EC_{50} 0.002 µmol TE/µmol peptide [42])
AEKTKIPAVF	f(73–82)	✓		✓	✓	-	-	-	
AEKTKIPA	f(73–80)	✓	✓	✓		-	-	-	
KIPAVF	f(77–82)				✓	-	-	-	
IPAVF	f(78–82)			✓		DPP-IV	44.7	Silveira, et al. [66]	

Table 1. *Cont.*

Peptide Sequence [1]	Region in Mature β-lg	Identified In				Reported Activity [2]	IC_{50}/EC_{50} Values [3]	Reference	References to Related Peptide Sequence [1,2,3,4]
		FPW_FF	FPW_ST	DBT_FF	DBT_ST				
VLDTDYKKY	f(94–102)			✓	✓	-	-	-	LDTDYKKYLLFCMENS (ABTS [67]), DTDYK (ABTS [56]), VLVLDTDYK (DPP-IVi IC_{50} 424.4 µM [66])
VLDTDYK	f(94–100)				✓	ABTS	na	Athira, et al. [56]	
						ACEi	946	Pihlanto-LeppÄLÄ, et al. [62]	
LDTDYKKY	f(95–102)			✓	✓	-	-	-	
LDTDYKK	f(95–101)			✓		ORAC, ACEi	na	Contreras, et al. [10]	
DTDYKKYLLF	f(96–105)			✓	✓	-	-	-	
DTDYKK	f(96–101)			✓	✓	-	-	-	
LVRTPEVDDEALEKF	f(123–135)			✓	✓	-	-	-	VRTPEVDDEALE, LVRTPEVDDEALE, RTPEVDDEALE (ABTS [56])
VRTPEVDDE	f(123–131)			✓	✓	ABTS	na	Athira, et al. [56]	
VRTPEVDDEALEK	f(123–134)			✓	✓	-	-	-	
RTPEVDDEALEK	f(124–134)			✓	✓	-	-	-	
TPEVDDEALEK	f(125–135)	✓	✓	✓	✓	DPP-IVi	319.5	Silveira, et al. [66]	
						ORAC	0.004	Power, et al. [42]	
						ABTS	na	Mann, et al. [57]	
ALKALPM	f(139–145)				✓	-	-	-	ALPMHIR (ACEi IC_{50} 43 µM [59], ORAC EC_{50} 0.035 µmol TE/µmol peptide [42])
KALPM	f(141–145)				✓	-	-	-	
ALPMH	f(142–146)		✓	✓	✓	ACEi	521	Mullally, et al. [59]	
						DPP-IVi	>100	Tulipano, et al. [68]	

[1] Peptide sequences presented with the one letter code. [2] ACEi: angiotensin I-converting enzyme inhibitory activity; DPP-IVi: dipeptidyl peptidase IV inhibitory activity; FRAP: ferric reducing antioxidant power; ORAC: oxygen radical antioxidant capacity. [3] IC_{50}: concentration of peptide resulting in 50% inhibition of ACE and DPP-IV activity reported as µM peptide; EC_{50}: half maximal effective concentration of peptide reported as µmol Trolox equivalent/µmol peptide; na: not applicable (no reported value). [4] DPPH: 2,2-diphenyl-1-picrylhydrazyl scavenging activity; ABTS: 2,2′-azino-bis-(3-ethylbenzothiazoline)-6-sulfonic acid radical scavenging activity.

Apart from the enzyme preparation used, the effect of hydrolysis conditions, ST vs. FF, for the same enzyme preparation was evident from the different peptides released from specific β-lg regions, e.g., f(15–22), f(73–82) and f(139–146) as shown in Figure 8. AEKTKIPAVF (β-lg f(73–82)) was present in DBT_FF and ST and appeared to act as an intermediate sequence for further hydrolysis by DBT. This, in turn, resulted in different peptides being released depending on the hydrolysis conditions, i.e., AEKTKIPA (β-lg f(73–80)) and IPAVF (β-lg f(78–82)) were identified in DBT_FF while KIPAVF (β-lg f(77–82)) was only found in DBT_ST (Table 1). This result indicates that hydrolysis conditions influenced the cleavage specificities for β-lg.

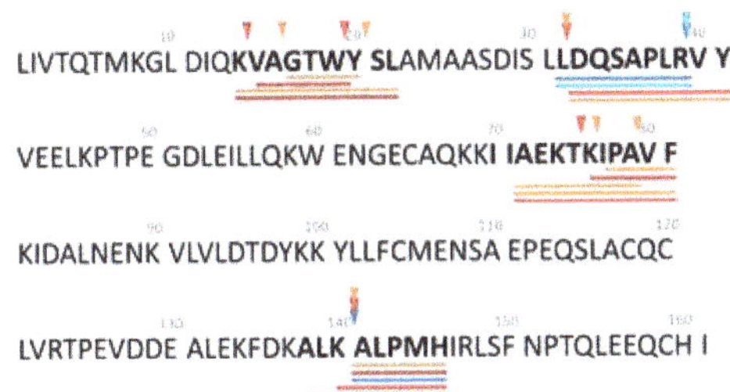

Figure 8. Primary sequence of bovine β-lactoglobulin showing specific cleavage sites and the peptides identified in whey protein concentrate hydrolysed using FlavorPro Whey (FPW) under pH-controlled (▼), non pH-controlled (▼) conditions and Debitrase (DBT) under pH-controlled (▼) and non pH-controlled (▼) conditions. Peptide sequences underlined in different colours indicate the peptides identified in the samples described above.

Similar observations can be made in the case of KVAGTWYSL (β-lg f(14–22)) which was detected in both DBT_FF and ST. Its derivatives, i.e., VAGTW (β-lg f(15–19)) and GTWYSL (β-lg f(17–22)), were present in DBT_ST and DBT_FF, respectively. Peptides related to β-lg f(14–22) have been reported to exert antioxidant properties, e.g., VAGTWY (β-lg f(15–20)) and VAGT (β-lg f(15–18)) showed ORAC values of 5.63 µmol TE/µmol peptide [42] and 1.66 µmol TE/mmol peptide [60], respectively. Furthermore, WYSL (β-lg f(19–22)) possessed DPPH and superoxide radical scavenging activity with EC_{50} values of 273.63 and 558.42 µM, respectively [61].

The lactokinin, ALPMHIR (β-lg f(142–148)), have been reported to exerted angiotensin I-converting enzyme (ACE) and dipeptidyl peptidase IV (DPP-IV) inhibitory and insulinotropic activities [59,68–71]. In addition, it has in vitro antioxidant activity with a reported ORAC value of 0.035 µmol TE/µmol peptide [42]. In the present study, ALPMH (β-lg f(142–146)) was detected in both DBT_ST and FF, whereas ALKALPM (β-lg f(139–145) and KALPM (β-lg f(141–145) were only detected in the DBT_ST WPH. In the case of FPW, only ALPMH was found in the FPW_ST WPH, while no lactokinin fragments (with >90% average local confidence) were detected in FPW_FF. The different peptides released from the WPHs generated using DBT and FPW under ST and FF conditions may explain the differences in antioxidant potencies of the resultant hydrolysates.

4. Conclusions

This study demonstrated the presence of antioxidant activity, using in vitro and cellular-based assays, in whey protein hydrolysates generated using Debitrase and FlavorPro Whey under ST and FF conditions. This appears to be the first report of the influence of enzymatic hydrolysis conditions on cellular antioxidant activity. The higher extent of hydrolysis in the DBT–WPHs may have contributed to more potent in vitro and cellular antioxidant properties when compared with the FPW–WPHs. The WPHs generated under ST conditions exerted stronger TEAC and ORAC activity. However, the antioxidant activity in the HepG2 cell-based assay was not influenced by the hydrolysis conditions used. This is despite the fact that differences in the peptides identified in the WPHs showed that hydrolysis conditions affected enzyme cleavage specificity.

Our findings extend the results of the previous studies by Le Maux et al. [12], Fernández and Kelly [23] and Butré et al. [26] showing the impact of hydrolysis conditions of whey proteins on the in vitro antioxidant activity, physicochemical properties and peptide profiles. The findings are relevant for the generation of whey protein derived antioxidant peptides at an industrial scale given that the hydrolysis conditions did not affect cellular antioxidant potencies. Nonetheless, the in vivo stability and bioavailability of the WPH-derived peptides remains to be established. Further investigations on the cellular antioxidant properties, specifically on the enzymes involved in oxidative stress as well as the immunomodulatory effects associated with various metabolic syndrome conditions, involving in vitro and in vivo studies are warranted.

Supplementary Materials: Supplementary data associated with this manuscript are available online at http://www.mdpi.com/2304-8158/9/5/582/s1, including **Figure S1:** Effect of different concentrations of oxidative stress inducers, 2,2'-azobis(2-amidinopropane) dihydrochloride (AAPH) and hydrogen peroxide (H_2O_2), on HepG2 cells viability. The results were expressed as the percentage of viable cells remaining following treatment with oxidative stress inducers compared to untreated control cells. Values represent mean ± SD ($n = 3$), **Figure S2:** Viability of HepG2 cells treated with the 4 h whey protein hydrolysates (WPHs) generated using FlavorPro Whey (FPW) and Debitrase (DBT) under pH-(ST) and non pH-controlled (FF) conditions. The results were expressed as the percentage of viable cells remaining following treatment with WPHs compared to untreated control cells. Values represent mean ± SD ($n = 3$).

Author Contributions: Conceptualisation, T.K. and R.J.F.; methodology, T.K., A.V.L.G. and B.D.; formal analysis, T.K., A.V.L.G. and B.D.; investigation, T.K., A.V.L.G. and B.D.; data curation, T.K., A.V.L.G. and B.D.; visualisation, T.K., A.V.L.G. and B.D.; writing—original draft preparation, T.K.; writing—review and editing, T.K., A.V.L.G., B.D., B.C. and R.J.F.; supervision, R.J.F. and B.C.; All authors have read and agreed to the submitted version of the manuscript.

Funding: This research was funded by Enterprise Ireland, under Dairy Processing Technology Centre (Grant number TC 2014 0016) for T.K. and the Department of Agriculture, Food, and the Marine (FIRM project 14/F/873) for A.V.L.G. Funding to A.V.L.G. and R.J.F. is gratefully acknowledged from the Irish Research Council 2018 Ulysses Programme project *'Food-grade protein hydrolysates from diverse origins targeting conditions of the metabolic syndrome (MetS): Assessment of their relevance for human health'* in conjunction with the University of Lille. Funding to B.D. and B.C. is gratefully acknowledged from Campus France PHC Ulysses 2018 *'Criblage des activités biologiques en relation avec le syndrome métabolique d'hydrolysats de protéines de diverses origines'* in conjunction with University of Limerick. This work was also supported by the Hauts-de-France region funding through the ALIBIOTECH research program. Some of the HPLC-MS analysis were performed on the REALCAT platform funded by the French National Research Agency (ANR) within the frame of the "Future Investments' program (ANR-11-EQPX-0037).

Acknowledgments: The authors would like to acknowledge Sara Paolella for LC-MS/MS of test samples and Alice Nongoneirma for helpful discussions.

Conflicts of Interest: The authors declare no conflict of interest.

References

1. Brandelli, A.; Daroit, D.J.; Corrêa, A.P.F. Whey as a source of peptides with remarkable biological activities. *Food Res. Int.* **2015**, *73*, 149–161. [CrossRef]
2. Nongonierma, A.B.; O'Keeffe, M.B.; FitzGerald, R.J. Milk protein hydrolysates and bioactive peptides. In *Advanced Dairy Chemistry: Volume 1b: Proteins: Applied Aspects*; McSweeney, P.L.H., O'Mahony, J.A., Eds.; Springer: New York, NY, USA, 2016; pp. 417–482.
3. Corrochano, A.R.; Buckin, V.; Kelly, P.M.; Giblin, L. Invited review: Whey proteins as antioxidants and promoters of cellular antioxidant pathways. *J. Dairy Sci.* **2018**, *101*, 4747–4761. [CrossRef] [PubMed]
4. Khan, I.T.; Nadeem, M.; Imran, M.; Ullah, R.; Ajmal, M.; Jaspal, M.H. Antioxidant properties of milk and dairy products: A comprehensive review of the current knowledge. *Lipids Health Dis.* **2019**, *18*, 41. [CrossRef] [PubMed]
5. Mangano, K.M.; Bao, Y.; Zhao, C. Nutritional properties of whey proteins. In *Whey Protein Production, Chemistry, Functionality, and Applications*; Guo, M., Ed.; John Wiley & Sons Ltd.: Hoboken, NJ, USA, 2019; pp. 103–140.
6. Poprac, P.; Jomova, K.; Simunkova, M.; Kollar, V.; Rhodes, C.J.; Valko, M. Targeting free radicals in oxidative stress-related human diseases. *Trends Pharm. Sci.* **2017**, *38*, 592–607. [CrossRef]

7. Niki, E. Assessment of antioxidant capacity in vitro and in vivo. *Free Radic. Biol. Med.* **2010**, *49*, 503–515. [CrossRef]

8. Udenigwe, C.C.; Aluko, R.E. Food protein-derived bioactive peptides: Production, processing, and potential health benefits. *J. Food Sci.* **2012**, *77*, R11–R24. [CrossRef]

9. Bamdad, F.; Bark, S.; Kwon, C.H.; Suh, J.-W.; Sunwoo, H. Anti-inflammatory and antioxidant properties of peptides released from β-lactoglobulin by high hydrostatic pressure-assisted enzymatic hydrolysis. *Molecules* **2017**, *22*, 949. [CrossRef]

10. Contreras, M.d.M.; Hernández-Ledesma, B.; Amigo, L.; Martín-Álvarez, P.J.; Recio, I. Production of antioxidant hydrolyzates from a whey protein concentrate with thermolysin: Optimization by response surface methodology. *LWT Food Sci. Technol.* **2011**, *44*, 9–15. [CrossRef]

11. Jiang, Z.; Yao, K.; Yuan, X.; Mu, Z.; Gao, Z.; Hou, J.; Jiang, L. Effects of ultrasound treatment on physicochemical, functional properties and antioxidant activity of whey protein isolate in the presence of calcium lactate. *J. Sci. Food Agric.* **2018**, *98*, 1522–1529. [CrossRef]

12. Le Maux, S.; Nongonierma, A.B.; Barre, C.; FitzGerald, R.J. Enzymatic generation of whey protein hydrolysates under pH-controlled and non pH-controlled conditions: Impact on physicochemical and bioactive properties. *Food Chem.* **2016**, *199*, 246–251. [CrossRef]

13. Madadlou, A.; Abbaspourrad, A. Bioactive whey peptide particles: An emerging class of nutraceutical carriers. *Crit. Rev. Food Sci. Nutr.* **2018**, *58*, 1468–1477. [CrossRef] [PubMed]

14. Neto, Y.A.A.H.; Rosa, J.C.; Cabral, H. Peptides with antioxidant properties identified from casein, whey, and egg albumin hydrolysates generated by two novel fungal proteases. *Prep. Biochem. Biotechnol.* **2019**, *49*, 639–648. [CrossRef] [PubMed]

15. O'Keeffe, M.B.; FitzGerald, R.J. Antioxidant effects of enzymatic hydrolysates of whey protein concentrate on cultured human endothelial cells. *Int. Dairy J.* **2014**, *36*, 128–135. [CrossRef]

16. Power, O.; Jakeman, P.; FitzGerald, R.J. Antioxidative peptides: Enzymatic production, in vitro and in vivo antioxidant activity and potential applications of milk-derived antioxidative peptides. *Amino Acids* **2013**, *44*, 797–820. [CrossRef] [PubMed]

17. Salami, M.; Moosavi-Movahedi, A.A.; Ehsani, M.R.; Yousefi, R.; Haertlé, T.; Chobert, J.-M.; Razavi, S.H.; Henrich, R.; Balalaie, S.; Ebadi, S.A.; et al. Improvement of the antimicrobial and antioxidant activities of camel and bovine whey proteins by limited proteolysis. *J. Agric. Food Chem.* **2010**, *58*, 3297–3302. [CrossRef]

18. Korhonen, H.; Pihlanto, A. Bioactive peptides: Production and functionality. *Int. Dairy J.* **2006**, *16*, 945–960. [CrossRef]

19. Tagliazucchi, D.; Helal, A.; Verzelloni, E.; Conte, A. Bovine milk antioxidant properties: Effect of in vitro digestion and identification of antioxidant compounds. *Dairy Sci. Technol.* **2016**, *96*, 657–676. [CrossRef]

20. De Carvalho, N.C.; Pessato, T.B.; Fernandes, L.G.R.; De Lima Zollner, R.; Netto, F.M. Physicochemical characteristics and antigenicity of whey protein hydrolysates obtained with and without pH control. *Int. Dairy J.* **2017**, *71*, 24–34. [CrossRef]

21. Cheison, S.C.; Kulozik, U. Impact of the environmental conditions and substrate pre-treatment on whey protein hydrolysis: A review. *Crit. Rev. Food Sci. Nutr.* **2017**, *57*, 418–453. [CrossRef]

22. Cheison, S.C.; Leeb, E.; Toro-Sierra, J.; Kulozik, U. Influence of hydrolysis temperature and pH on the selective hydrolysis of whey proteins by trypsin and potential recovery of native alpha-lactalbumin. *Int. Dairy J.* **2011**, *21*, 166–171. [CrossRef]

23. Fernández, A.; Kelly, P. pH-stat vs. Free-fall pH techniques in the enzymatic hydrolysis of whey proteins. *Food Chem.* **2016**, *199*, 409–415. [CrossRef] [PubMed]

24. Butré, C.I.; Sforza, S.; Wierenga, P.A.; Gruppen, H. Determination of the influence of the pH of hydrolysis on enzyme selectivity of *Bacillus licheniformis* protease towards whey protein isolate. *Int. Dairy J.* **2015**, *44*, 44–53. [CrossRef]

25. Butré, C.I.; Wierenga, P.A.; Gruppen, H. Effects of ionic strength on the enzymatic hydrolysis of diluted and concentrated whey protein isolate. *J. Agric. Food Chem.* **2012**, *60*, 5644–5651. [CrossRef] [PubMed]

26. Butré, C.I.; Sforza, S.; Gruppen, H.; Wierenga, P.A. Introducing enzyme selectivity: A quantitative parameter to describe enzymatic protein hydrolysis. *Anal. Bioanal. Chem.* **2014**, *406*, 5827–5841. [CrossRef] [PubMed]

27. Apak, R.; Özyürek, M.; Güçlü, K.; Çapanoğlu, E. Antioxidant activity/capacity measurement. 2. Hydrogen atom transfer (HAT)-based, mixed-mode (electron transfer (ET)/HAT), and lipid peroxidation assays. *J. Agric. Food Chem.* **2016**, *64*, 1028–1045. [CrossRef] [PubMed]

28. Spellman, D.; McEvoy, E.; O'Cuinn, G.; FitzGerald, R.J. Proteinase and exopeptidase hydrolysis of whey protein: Comparison of the TNBS, OPA and pH stat methods for quantification of degree of hydrolysis. *Int. Dairy. J.* **2003**, *13*, 447–453. [CrossRef]

29. O'Loughlin, I.B.; Murray, B.A.; Kelly, P.M.; FitzGerald, R.J.; Brodkorb, A. Enzymatic hydrolysis of heat-induced aggregates of whey protein isolate. *J. Agric. Food Chem.* **2012**, *60*, 4895–4904. [CrossRef]

30. Spellman, D.; O'Cuinn, G.; FitzGerald, R.J. Bitterness in *Bacillus* proteinase hydrolysates of whey proteins. *Food Chem.* **2009**, *114*, 440–446. [CrossRef]

31. Le Maux, S.; Nongonierma, A.B.; Lardeux, C.; FitzGerald, R.J. Impact of enzyme inactivation conditions during the generation of whey protein hydrolysates on their physicochemical and bioactive properties. *Int. J. Food Sci. Technol.* **2018**, *53*, 219–227. [CrossRef]

32. Re, R.; Pellegrini, N.; Proteggente, A.; Pannala, A.; Yang, M.; Rice-Evans, C. Antioxidant activity applying an improved ABTS radical cation decolorization assay. *Free Radic. Biol. Med.* **1999**, *26*, 1231–1237. [CrossRef]

33. Yarnpakdee, S.; Benjakul, S.; Kristinsson, H.G.; Bakken, H.E. Preventive effect of nile tilapia hydrolysate against oxidative damage of HepG2 cells and DNA mediated by H_2O_2 and AAPH. *J. Food Sci. Technol.* **2015**, *52*, 6194–6205. [CrossRef] [PubMed]

34. Nongonierma, A.B.; Cadamuro, C.; Le Gouic, A.; Mudgil, P.; Maqsood, S.; FitzGerald, R.J. Dipeptidyl peptidase IV (DPP-IV) inhibitory properties of a camel whey protein enriched hydrolysate preparation. *Food Chem.* **2019**, *279*, 70–79. [CrossRef] [PubMed]

35. O'Keeffe, M.B.; FitzGerald, R.J. Identification of short peptide sequences in complex milk protein hydrolysates. *Food Chem.* **2015**, *184*, 140–146. [CrossRef] [PubMed]

36. Heberle, H.; Meirelles, G.V.; Da Silva, F.R.; Telles, G.P.; Minghim, R. InteractiVenn: A web-based tool for the analysis of sets through Venn diagrams. *BMC Bioinform.* **2015**, *16*, 169. [CrossRef]

37. Haileselassie, S.S.; Lee, B.H.; Gibbs, B.F. Purification and identification of potentially bioactive peptides from enzyme-modified cheese. *J. Dairy Sci.* **1999**, *82*, 1612–1617. [CrossRef]

38. Nongonierma, A.B.; FitzGerald, R.J. Enzymes exogenous to milk in dairy technology | Proteinases. In *Encyclopedia of Dairy Sciences*, 2nd ed.; Fuquay, J.W., Ed.; Academic Press: San Diego, CA, USA, 2011; pp. 289–296.

39. Spellman, D.; O'Cuinn, G.; FitzGerald, R.J. Physicochemical and sensory characteristics of whey protein hydrolysates generated at different total solids levels. *J. Dairy Res.* **2005**, *72*, 138–143. [CrossRef]

40. O'Sullivan, D.; Nongonierma, A.B.; FitzGerald, R.J. Bitterness in sodium caseinate hydrolysates: Role of enzyme preparation and degree of hydrolysis. *J. Sci. Food Agric.* **2017**, *97*, 4652–4655. [CrossRef]

41. Kalyankar, P.; Zhu, Y.; O'Cuinn, G.; FitzGerald, R.J. Investigation of the substrate specificity of glutamyl endopeptidase using purified bovine β-casein and synthetic peptides. *J. Agric. Food Chem.* **2013**, *61*, 3193–3204. [CrossRef]

42. Power, O.; Fernández, A.; Norris, R.; Riera, F.A.; FitzGerald, R.J. Selective enrichment of bioactive properties during ultrafiltration of a tryptic digest of β-lactoglobulin. *J. Funct. Foods* **2014**, *9*, 38–47. [CrossRef]

43. Giblin, L.; Yalçın, A.S.; Biçim, G.; Krämer, A.C.; Chen, Z.; Callanan, M.J.; Arranz, E.; Davies, M.J. Whey proteins: Targets of oxidation, or mediators of redox protection. *Free Radic. Res.* **2019**, *53*, 1136–1152. [CrossRef]

44. Xu, R.; Liu, N.; Xu, X.; Kong, B. Antioxidative effects of whey protein on peroxide-induced cytotoxicity. *J Dairy Sci.* **2011**, *94*, 3739–3746. [CrossRef] [PubMed]

45. Corrochano, A.R.; Ferraretto, A.; Arranz, E.; Stuknytė, M.; Bottani, M.; O'Connor, P.M.; Kelly, P.M.; De Noni, I.; Buckin, V.; Giblin, L. Bovine whey peptides transit the intestinal barrier to reduce oxidative stress in muscle cells. *Food Chem.* **2019**, *288*, 306–314. [CrossRef] [PubMed]

46. Kong, B.; Peng, X.; Xiong, Y.L.; Zhao, X. Protection of lung fibroblast MRC-5 cells against hydrogen peroxide-induced oxidative damage by 0.1–2.8 kDa antioxidative peptides isolated from whey protein hydrolysate. *Food Chem.* **2012**, *135*, 540–547. [CrossRef] [PubMed]

47. Zhang, Q.-X.; Ling, Y.-F.; Sun, Z.; Zhang, L.; Yu, H.-X.; Kamau, S.M.; Lu, R.-R. Protective effect of whey protein hydrolysates against hydrogen peroxide-induced oxidative stress on PC12 cells. *Biotechnol. Lett.* **2012**, *34*, 2001–2006. [CrossRef] [PubMed]

48. Piccolomini, A.; Iskandar, M.; Lands, L.; Kubow, S. High hydrostatic pressure pre-treatment of whey proteins enhances whey protein hydrolysate inhibition of oxidative stress and IL-8 secretion in intestinal epithelial cells. *Food Nutr. Res.* **2012**, *56*, 17549. [CrossRef] [PubMed]

49. Iskandar, M.; Lands, L.; Sabally, K.; Azadi, B.; Meehan, B.; Mawji, N.; Skinner, C.; Kubow, S. High hydrostatic pressure pretreatment of whey protein isolates improves their digestibility and antioxidant capacity. *Foods* **2015**, *4*, 184–207. [CrossRef]

50. Pyo, M.C.; Yang, S.-Y.; Chun, S.-H.; Oh, N.S.; Lee, K.-W. Protective effects of maillard reaction products of whey protein concentrate against oxidative stress through an Nrf2-dependent pathway in HepG2 cells. *Biol. Pharm. Bull.* **2016**, *39*, 1437–1447. [CrossRef]

51. Kim, G.-N.; Kwon, Y.-I.; Jang, H.-D. Protective mechanism of quercetin and rutin on 2,2′-azobis(2-amidinopropane)dihydrochloride or Cu^{2+}-induced oxidative stress in HepG2 cells. *Toxicol. In Vitro* **2011**, *25*, 138–144. [CrossRef]

52. Honzel, D.; Carter, S.G.; Redman, K.A.; Schauss, A.G.; Endres, J.R.; Jensen, G.S. Comparison of chemical and cell-based antioxidant methods for evaluation of foods and natural products: Generating multifaceted data by parallel testing using erythrocytes and polymorphonuclear cells. *J. Agric. Food Chem.* **2008**, *56*, 8319–8325. [CrossRef]

53. Lima, C.F.; Fernandes-Ferreira, M.; Pereira-Wilson, C. Phenolic compounds protect HepG2 cells from oxidative damage: Relevance of glutathione levels. *Life Sci.* **2006**, *79*, 2056–2068. [CrossRef]

54. Athira, S.; Mann, B.; Sharma, R.; Kumar, R. Ameliorative potential of whey protein hydrolysate against paracetamol-induced oxidative stress. *J. Dairy Sci.* **2013**, *96*, 1431–1437. [CrossRef] [PubMed]

55. Ashok, N.R.; Vivek, K.H.; Aparna, H.S. Antioxidative role of buffalo (*Bubalus bubalis*) colostrum whey derived peptides during oxidative damage. *Int. J. Pept. Res.* **2019**, *25*, 1501–1508. [CrossRef]

56. Athira, S.; Mann, B.; Saini, P.; Sharma, R.; Kumar, R.; Singh, A.K. Production and characterisation of whey protein hydrolysate having antioxidant activity from cheese whey. *J. Sci. Food Agric.* **2015**, *95*, 2908–2915. [CrossRef] [PubMed]

57. Mann, B.; Kumari, A.; Kumar, R.; Sharma, R.; Prajapati, K.; Mahboob, S.; Athira, S. Antioxidant activity of whey protein hydrolysates in milk beverage system. *J. Food Sci. Technol.* **2015**, *52*, 3235–3241. [CrossRef]

58. Picariello, G.; Iacomino, G.; Mamone, G.; Ferranti, P.; Fierro, O.; Gianfrani, C.; Di Luccia, A.; Addeo, F. Transport across Caco-2 monolayers of peptides arising from *in vitro* digestion of bovine milk proteins. *Food Chem.* **2013**, *139*, 203–212. [CrossRef]

59. Mullally, M.M.; Meisel, H.; FitzGerald, R.J. Identification of a novel angiotensin-I-converting enzyme inhibitory peptide corresponding to a tryptic fragment of bovine β-lactoglobulin. *FEBS Lett.* **1997**, *402*, 99–101. [CrossRef]

60. O'Keeffe, M.B.; Conesa, C.; FitzGerald, R.J. Identification of angiotensin converting enzyme inhibitory and antioxidant peptides in a whey protein concentrate hydrolysate produced at semi-pilot scale. *Int. J. Food Sci. Technol.* **2017**, *52*, 1751–1759. [CrossRef]

61. Hernández-Ledesma, B.; Miguel, M.; Amigo, L.; Aleixandre, M.A.; Recio, I. Effect of simulated gastrointestinal digestion on the antihypertensive properties of synthetic β-lactoglobulin peptide sequences. *J. Dairy Res.* **2007**, *74*, 336–339. [CrossRef]

62. Pihlanto-LeppÄLÄ, A.; Koskinen, P.; Piilola, K.; Tupasela, T.; Korhonen, H. Angiotensin I-converting enzyme inhibitory properties of whey protein digests: Concentration and characterization of active peptides. *J. Dairy Res.* **2000**, *67*, 53–64. [CrossRef]

63. Pellegrini, A.; Dettling, C.; Thomas, U.; Hunziker, P. Isolation and characterization of four bactericidal domains in the bovine β-lactoglobulin. *Biochim. Biophys. Acta (BBA) Gen. Subj.* **2001**, *1526*, 131–140. [CrossRef]

64. Tavares, T.; Contreras, M.d.M.; Amorim, M.; Pintado, M.; Recio, I.; Malcata, F.X. Novel whey-derived peptides with inhibitory effect against angiotensin-converting enzyme: In vitro effect and stability to gastrointestinal enzymes. *Peptides* **2011**, *32*, 1013–1019. [CrossRef] [PubMed]

65. Lacroix, I.M.E.; Li-Chan, E.C.Y. Overview of food products and dietary constituents with antidiabetic properties and their putative mechanisms of action: A natural approach to complement pharmacotherapy in the management of diabetes. *Mol. Nutr. Food Res.* **2014**, *58*, 61–78. [CrossRef] [PubMed]

66. Silveira, H.; Moraes, H.; Oliveira, N.; Coutinho, E.S.F.; Laks, J.; Deslandes, A. Physical exercise and clinically depressed patients: A systematic review and meta-analysis. *Neuropsychobiology* **2013**, *67*, 61–68. [CrossRef] [PubMed]

67. Bertucci, J.I.; Liggieri, C.S.; Colombo, M.L.; Vairo Cavalli, S.E.; Bruno, M.A. Application of peptidases from *Maclura pomifera* fruit for the production of active biopeptides from whey protein. *LWT Food Sci. Technol.* **2015**, *64*, 157–163. [CrossRef]
68. Tulipano, G.; Sibilia, V.; Caroli, A.M.; Cocchi, D. Whey proteins as source of dipeptidyl dipeptidase IV (dipeptidyl peptidase-4) inhibitors. *Peptides* **2011**, *32*, 835–838. [CrossRef]
69. Tulipano, G.; Faggi, L.; Nardone, A.; Cocchi, D.; Caroli, A.M. Characterisation of the potential of β-lactoglobulin and α-lactalbumin as sources of bioactive peptides affecting incretin function: *In silico* and in vitro comparative studies. *Int. Dairy J.* **2015**, *48*, 66–72. [CrossRef]
70. FitzGerald, R.; Meisel, H. Lactokinins: Whey protein-derived ACE inhibitory peptides. *Food/Nahrung* **1999**, *43*, 165–167. [CrossRef]
71. Maes, W.; Van Camp, J.; Vermeirssen, V.; Hemeryck, M.; Ketelslegers, J.M.; Schrezenmeir, J.; Van Oostveldt, P.; Huyghebaert, A. Influence of the lactokinin Ala-Leu-Pro-Met-His-Ile-Arg (ALPMHIR) on the release of endothelin-1 by endothelial cells. *Regul. Pept.* **2004**, *118*, 105–109. [CrossRef]

Article

Antioxidant and Anti-Apoptotic Properties of Oat Bran Protein Hydrolysates in Stressed Hepatic Cells

Ramak Esfandi [1], William G. Willmore [2,3] and Apollinaire Tsopmo [1,3,*]

[1]   Food Science and Nutrition Program, Department of Chemistry, Carleton University, 1125 Colonel By Drive, Ottawa, ON K1S 5B6, Canada; RamakEsfandi@cmail.carleton.ca

[2]   Department of Biology, Carleton University, 1125 Colonel By Drive, Ottawa, ON K1S 5B6, Canada; Bill_Willmore@carleton.ca

[3]   Institute of Biochemistry, Carleton University, 1125 Colonel By Drive, Ottawa, ON K1S 5B6, Canada

*   Correspondence: apollinaire_tsopmo@carleton.ca; Tel.: +1-613-520-2600 (ext. 3122)

Received: 25 March 2019; Accepted: 9 May 2019; Published: 11 May 2019

Abstract: The objective of this work was to find out how the method to extract proteins and subsequent enzymatic hydrolysis affect the ability of hepatic cells to resist oxidative stress. Proteins were isolated from oat brans in the presence of Cellulase (CPI) or Viscozyme (VPI). Four protein hydrolysates were produced from CPI and four others from VPI when they treated with Alcalase, Flavourzyme, Papain, or Protamex. Apart from CPI-Papain that reduced the viability of cell by 20%, no other hydrolysate was cytotoxic in the hepatic HepG2 cells. In the cytoprotection test, VPI-Papain and VPI-Flavourzyme fully prevented the damage due to peroxyl radical while CPI-Papain and CPI-Alcalase enhanced the cellular damage. Cells treated with VPI-hydrolysates reduced intracellular reactive oxygen species (ROS) by 20–40% and, also increased the intracellular concentration of glutathione, compared to CPI-hydrolysates. In antioxidant enzyme assays, although all hydrolysates enhanced the activity of both superoxide dismutase and catalase by up to 2- and 3.4-fold, respectively relative the control cells, the largest increase was due to VPI-Papain and VPI-Flavourzyme hydrolysates. In caspase-3 assays, hydrolysates with reduced ROS or enhanced antioxidant enzyme activities were able to reduce the activity of the pro-apoptotic enzyme, caspase-3 indicating that they prevented oxidative stress-induced cell death.

Keywords: food peptides; reactive oxygen species; antioxidant enzymes

1. Introduction

The liver as a main detoxifying organ is prone to oxidative stress because of continuous exposure to reactive oxygen species (ROS) and toxicants. The production of ROS beyond the ability of cells to neutralize the radicals may cause damage to lipids, proteins and nucleotides which eventually can initiate liver-related injuries and diseases [1]. Cellular defence systems include small molecules and enzymes that are used to maintain a balance between oxidation and reduction. However, ageing, constant exposure to environmental toxicants or drugs can shift the balance towards a greater accumulation of oxidants (e.g., free radicals). It then becomes important to provide cells with additional amounts of antioxidants in the form of the supplement of formulated food products.

Antioxidant molecules have been considered as a strategy to prevent or reduce the incidence of many health-related conditions including liver diseases in which oxidative stress is present. Although there are synthetic antioxidants, natural ones have the advantage of being safer as well as being, in certain cases, multifunctional (e.g., antioxidant, anti-inflammatory, anti-apoptotic) [2–4]. Polyphenols have been extensively studied for their hepatic protective effect in animal and cell culture models, and this is summarized in a recent review [5]. The protection is often evaluated and quantified by electron donating compounds (e.g., vitamins E and C, glutathione) or antioxidant enzymes such as peroxidases and superoxide dismutase. In recent years, there has been an interest in the determination of the biological activity of hydrolyzed food proteins including hepato-protective effects. The hydrolyzed proteins contain peptides which are believed to be safer, and many are multifunctional [6]. In recent works, hydrolyzed corn proteins showed a hepatoprotective effect against carbon tetrachloride-induced liver injury in mice, characterized by a reduction in the oxidation of lipids, and an increase in the activity of superoxide dismutase and glutathione concentrations [7]. In hepatocyte cells, hydrolyzed rice bran proteins increased the intracellular glutathione concentrations by 2-fold in oxidatively stressed cells due to increased expression of γ-glutamylcysteine synthetase [8] while antioxidant peptides from microalgae had protection against alcohol-induced damage in HepG2/CYP2E1 cells [9].

Oat grains have the highest amount of proteins (up to 17% weight) amongst cereals most (50–80%) of which are globulins. Previous works on oat proteins resulted in the production of hydrolysates with radical scavenging activities [10,11], metal (calcium, iron) binding and inhibition of linoleic acid oxidation [12]. In one of those works showed that proteins extracted from oat brans with the aid of polysaccharide degrading enzymes, Cellulase and Viscozyme, had different susceptibility when hydrolyzed with Alcalase, Flavourzyme, Papain, and Protamex [13]. In addition, the extraction procedure also affected free radical scavenging and metal binding capacities of hydrolysates produced with the same protease. As a follow-up, this work aimed to determine at a cellular level, the potential of the hydrolysates to prevent oxidative stress and apoptosis in hepatic HepG2 cells.

2. Materials and Methods

2.1. Chemicals and Reagents

Caspase-3/CPP32 colorimetric assay Kit (Biovision, Catalog# K106-100) (# K106-100) was purchased from Biovision (Mountain View, CA, USA). Enzymes (glutathione peroxidase, catalase, copper-zinc superoxide dismutase, xanthine oxidase, glutathione reductase (GR)), 3-(4,5-dimethylthiazol-2-yl)-2,5-diphenyltetrazolium bromide (MTT), reduced glutathione (GSH), oxidized glutathione (GSSG), nicotinamide adenine dinucleotide phosphate (NADPH), 4-(2-hydroxyethyl)-1-piperazineethanesulfonic acid (HEPES), dihydrochlorofluorescein diacetate (DCFH$_2$-DA), bathocuproinedisulfonic acid disodium salt (BCS), nitroblue tetrazolium chloride (NBT), diethylenetriaminepentaacetic acid (DETAPAC), xanthine, 5,5-dithiobis-(2-nitrobenzoic acid) (DTNB), triethanolamine, 2-vinylpyridine were obtained from Sigma Aldrich (Oakville, ON, Canada). Hydrogen peroxide (H$_2$O$_2$), 2,2′-azobis(2-amidinopropane) dihydrochloride (AAPH), sodium azide (NaN$_3$), ethylenediaminetetraacetic acid disodium (EDTA), antibiotic-antimycotic (100×) were obtained from Fisher Scientific (Ottawa, ON, Canada).

2.2. Oat Bran Protein Hydrolysates

Oat protein isolates and their hydrolysates used in this study are from recent work [13]. Briefly, polysaccharides in defatted brans were cleaved (pH 4.5, 45 °C, 90 min) with either Cellulase (20 units/g) or Viscozyme (3 units/g). The Cellulase Protein Isolate (CPI) and the Viscozyme Protein Isolate (VPI) were subsequently obtained after solubilisation at pH 9.5 and precipitation at isoelectric point (pH 4.5). Each protein isolate was digested with proteolytic enzymes (Alcalase, Flavourzyme, Papain and Protamex), to produce eight protein hydrolysates. The CPI-derived hydrolysates were named CPI-Al, CPI-Fl, CPI-Pa, and CPI-Pr, respectively while the ones from VPI were named VPI-Al, VPI-Fl, VPI-Pa, and VPI-Pr, respectively. Detailed preparation methods and protein contents are as provided in the literature [13].

2.3. Cell Culture, Cytotoxicity and Cytoprotective Experiments

Human hepatic carcinoma HepG2 cells (ATCC® HB-8065™) were obtained from Cedarlane Laboratories Ltd (Burlington, ON, Canada). They were grown in Dulbecco's Modified Eagle Media (DMEM) with 10% fetal bovine serum (FBS, from Wisent Bioproducts St-Bruno, QC, Canada). Cells were passaged every 3–4 days and maintained in a humidified incubator at 37 °C supplemented with 5% CO_2-95% air.

The cytotoxicity and cytoprotection effects of protein hydrolysates were evaluated using a modified MTT assay described by Nair and Liu, 2010 [14]. The concentration of protein hydrolysates was 100 µg/mL as previous work on other hydrolysates found no toxic effect up to 1 mg/mL [15]. A volume of 200 µL cells at a density of 2×10^4 cells/mL (in DMEM containing 10% FBS and 1% antibiotic) was transferred to clear 96-well tissue culture plates and incubated for 24 h. Media was then removed, and the cells washed with phosphate buffer saline solution (PBS, pH 7.2) followed by 24 h incubation with protein hydrolysates and another wash. Subsequently, 200 µL of media was added to the cells intended for cytotoxicity studies while 200 µL of media containing 10 mM of AAPH were added to those used for cytoprotection evaluation. After another 24 h incubation, cells were washed twice with PBS. The determination of viable cells in each case was initiated by adding 100 µL of media and 50 µL MTT (1 mg/mL) were added to all wells and maintained for 1 h at 37 °C. The MTT solution was replaced with 50 µL of DMSO before absorbance reading (570 nm) with background subtraction (630 nm). The cytotoxicity effect of proteins hydrolysates was compared to normal cells (i.e., no treatment), the cytoprotection was to both normal, and AAPH treated cells.

2.4. Determination of Intracellular Reactive Oxygen Species

Reactive oxygen species (ROS) scavenging activities of protein hydrolysates were determined based on a reported procedure [16]. Briefly, HepG2 Cells were seeded at 4×10^4 cell/mL (200 µL/well) in a dark 96-well tissue culture plate and incubated for 24 h prior after which they were washed with PBS (2×200 µL, pH 7.4). Media with or without protein hydrolysates were added to the wells and allowed to interact for 24 h. Cells were washed again with PBS followed by addition of AAPH (10 mM), interaction over 45 min and another wash. To determine ROS, 200 µL of 40 mM $DCFH_2$-DA in 10 mM HEPES buffer was added to the wells. Fluorescent intensity was immediately measured over 60 min with 5 min intervals for one hour, using Fluostar Optima, (BMG Labtech, Offenburg, Germany) at an excitation wavelength at 485 nm and emission at 530 nm. Untreated cells were used as the negative control (NEG), whereas the positive control (POS) consists of cells treated with AAPH only. ROS was calculated according to the equation below and normalized to protein contents from the Lowry assay.

$$\text{ROS (\%)} = (((\text{Final reading} - \text{Initial reading})/\text{Initial reading}) \times 100)/(\text{mg protein/mL}). \tag{1}$$

2.5. Preparation of Cell Extracts for Glutathione and Enzyme Assays

HepG2 cells were seeded at 5×10^5 cell/plate in 60 mm tissue culture plates and incubated at 37 °C for 24 h after which media was discarded, and the plates washed with PBS (2×4 mL, pH 7.4). Hydrolyzed proteins (4 mL in media) were added to plates and incubated for 24 h. Cells were washed with PBS and treated with 10 mM AAPH (4 mL in media) for 24 h. Cells were detached with 0.25% trypsin (0.5 mL) over 5 min at 37 °C, inhibited with 1 mL culture media, transferred to 1.5 mL vials. The suspension was centrifuged at $1000 \times g$ for 5 min at 4 °C. Cells were collected and re-suspended in 1 mL ice-cold potassium phosphate buffer (0.1 M, pH 7.5) containing 5 mM EDTA, 0.1% Triton X, and 0.6% sulfosalicylic acid to be lysed by sonication for determination of antioxidant enzyme activities or in the lysis buffer provided in the Caspase-3 assay kit. The lysis on ice used a probe-type sonicator (Vibra-Cell, Sonics & Materials Inc., Newtown, CT, USA) with cycles set at 1 min for 15 s on and 10 s off. For glutathione assay, 1 mL of 5% ice-cold sulfosalicylic acid bubbled with 100% nitrogen was used for cell lysis. Cell extracts (supernatants) were obtained by centrifugation at $13,000 \times g$ for 5 min and immediately used for assays.

2.6. Glutathione Assay

Total and oxidized glutathione were measured based on a method described by Rahman, et al. [17]. Total glutathione measurement was done by adding to a 96-well clear microplate, 20 µL of cell extracts,

GSH standard (0.103–26.4 μM), or buffer (for blank) and an equal volume (60 μL) of 0.8 mM DTNB and 100 IU/mL of glutathione reductase. The reaction rate between GSH in the lysate and DTNB were recorded at 412 every 20 s for 2 min after the addition of 60 μL of 0.8 mM β-NADPH. In the case of oxidized glutathione, 2 μL of 2-vinylpyridine (0.185 mM) were added to 100 μL of cell extract, GSSG standard (0.103–26.4 μM), or blank in 1.5 mL microcentrifuge tubes, mixed and incubated for 1 h at room temperature. Then, 6 μL of 6-times diluted triethanolamine were added to the tubes and vortex mixed. After 10 min incubation time, the next steps were as above for total glutathione. Concentrations were calculated based on a standard curve of glutathione normalized to protein content.

2.7. Antioxidant Enzyme Assays

2.7.1. Catalase

Catalase activity was measured using a method described by Beers and Sizer [18]. Cell extracts 200 μL) were mixed with 1790 μL of potassium phosphate buffer (0.05 M, pH 7.0) in UV disposable cuvettes. Ten microliters of 30% H_2O_2 were added followed by measurement of its decomposition are at 240 nm using a Cary 50 Bio UV-Vis spectrophotometer with 18-cell changer (Varian Inc., Mississauga, ON, Canada). The rate was used to calculate catalase activity and expressed as a percentage of the control.

2.7.2. Glutathione Peroxidase

The activity of glutathione peroxidase (GPx) was measured based on previous literature [19]. A potassium phosphate stock buffer (0.05 M, pH 7.0) containing 1.1 mM EDTA and 1.1 mM NaN_3 was used to make to GSH (1 mM), and 1 unit/mL glutathione reductase (1 IU/mL) before the assay and NADPH (4 mM) that was kept on ice. Into a clear 96 well-plate, 187.5 μL of the above buffer was added followed by 12.5 μL of NADPH and 25 μL of either cell lysate or GPx standards (0.02–0.1 U/mL) or buffer for the blank. The reaction was initiated by the addition of 2.5 mM H_2O_2 and incubated at 30 °C for 5 min then, the change in the absorbance was measured at 340 nm due to the disappearance of NADPH over 4 min was used to calculate the activity as Units of GPx activity/mg protein.

2.7.3. Superoxide Dismutase

The activity of superoxide dismutase was measured using a procedure described by Spitz and Oberley with modifications [20]. The assay solution was 50 mM phosphate buffer (pH 7.8) containing DETAPAC (1.25 mM), BSA (0.16 mg/mL), Catalase (1.25 U/mL), NBT (70 μM), xanthine (125 mM), and BCS (62.5 μM). Xanthine oxidase 0.06 U/mL was made in 1.34 mM DETAPAC buffer. Cell extracts were diluted to eight concentrations (2–500 mg/mL) while CuZnSOD standard concentrations were 2–500 ng/mL. Specifically, 200 μL of the assay solution and 25 μL of sample or standard were added to a clear 96-well microplate. The reaction then initiated by the addition of 25 μL xanthine oxidase followed by kinetic measurement every 15 s for 3 min and at 560 nm. Rates of reduced of NBT were used to determine concentrations of cell lysate that inhibited 50% of the activity of SOD enzyme.

2.8. Caspase-3 Assay

Caspase-3/CPP32 Colorimetric Assay Kit (Biovision, Catalog #K 106-100) was used to determine the apoptosis induced by AAPH and with or without pre-treatment of cells with protein hydrolysates. The analysis was done according to the manufacturer's instructions. In summary, cell lysates were diluted to obtain 50 μg protein/mL. Then, 50 μL of diluted lysate was mixed with 50 μL of reaction buffer containing 10 mM DTT and 5 μL of 4 mM Asp-Glu-Val-Asp *p*-nitroanilide (DEVD-pNA) substrate. Tubes were incubated at 37 °C for 2 h followed by measurement of the absorbance at 405 nm. A blank was also prepared using cell lysis buffer only.

2.9. Statistical Analysis

All experiments were performed in triplicate (n = 3), and one-way analysis of variance (ANOVA) tests (SAS® Software, SAS OnDemand, 9.4, 2017, SAS Institute Inc., Cary, NC, USA) were used to compare mean values. Significant differences were evaluated using Duncan's multiple range test ($p < 0.05$).

3. Results and Discussion

3.1. Cytotoxicity and Cytoprotection

The reduction of the MTT tetrazolium salt to formazan, by mitochondrial oxidoreductase enzymes, was used to determine the cytotoxicity and cytoprotection of hydrolyzed oat proteins. The concentration of formazan is proportional to the number of viable cells. In the cytotoxicity test (Figure 1A), exposure of HepG2 cells to protein hydrolysates showed that only CPI-Pa significantly reduced the number of viable cells. CPI-Pa reduction was about 21%. Four hydrolysates reduce the viability by about 7% ($p < 0.05$) while three others increased the viability by 11–15 % ($p > 0.05$). One of the non-hydrolyzed proteins, CPI, did not affect cell growth while the other, VPI, significantly enhance the growth by about 40%. Cell growth promotion has been reported using food protein hydrolysates. In one study, for example, soy protein hydrolysate enhanced the growth of mouse hybridoma ME-750 cells by 7% while the increase was 48% for wheat gluten hydrolysate [21]. The effect of protein hydrolysates then appeared to vary depending on the source of proteins, the protein extraction procedure and the specificity of the proteases used. Similar to CPI-Pa, an Alcalase gelatin hydrolysate was reported to decrease the viability of HepG2 cells by 20% at 200 μg/mL [22]. The beneficial effect of hydrolyzed proteins of cells may come from their nutritive or regulatory effects but also, potentially from their action on maintaining the redox balance.

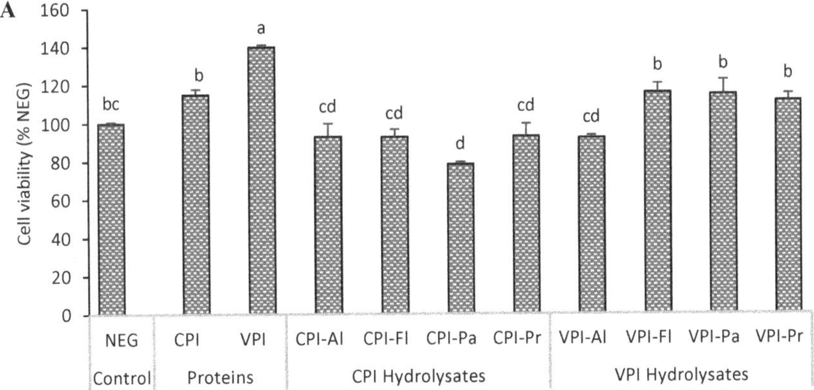

Figure 1. *Cont.*

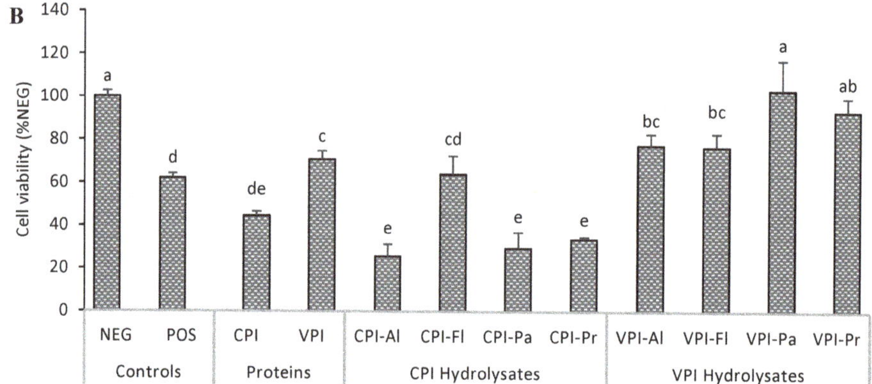

Figure 1. Cytotoxicity (**A**) and cytoprotective (**B**) effects of oat bran protein hydrolysates, digested with Cellulase (CPI) or with Viscozyme (VPI) and their hydrolysates produced with Alcalase (Al), Flavourzyme (Fl), Papain (Pa) or Protamex (Pr) in HepG2 cells. Cells were exposed to hydrolysate (100 µg/mL) for 24 h (**A**) or exposed to hydrolysates for 24 h followed by AAPH for 24 h (**A**) before the determination of viable cells. NEG: normal cells; POS: cells treated with AAPH alone. Data with different letters showed significant difference ($p < 0.05$) from Duncan's multiple range test ($n = 3$).

The exposure of HepG2 cells to AAPH radical caused a significant decrease in the viability of cells (Figure 1B), and this is consistent with previous works on oxidatively-induced cellular damage [23,24]. AAPH induced oxidative stress reduced cell viability to 62.1% ± 2.1% relative to normal control cells. When the cells were pre-treated with hydrolysate before being stressed, different effects were found. VPI-Pa and VPI-Pr caused no reduction in viability, relative to normal cells, and was, therefore, the most cytoprotective. The other protective hydrolysates were VPI-Al and VPI-Fl. CPI-FL did not affect the oxidant activity (i.e., AAPH) while the remaining three hydrolysates caused a decrease in viability (Figure 1B).

It appeared that hydrolysates from Viscozyme extracted proteins (VPI) had better cytoprotection than those derived from Cellulase extracted proteins (CPI). Previous work found that the composition and the concentration of polypeptides in CPI and VPI were different and this translated in different peptides composition hydrolysates [13]. That work identified 92, 171 and 609 peptides made of 8–26 amino acids in VPI-Pa, CPI-Pa, and VPI-Pr, respectively. The activity of the hydrolysate is related to several factors that may include the number of peptides, their size, charge or hydrophobicity and the overall sequences.

3.2. Determination of Intracellular Reactive Species and Glutathione

The scavenging activities of the protein hydrolysates against AAPH-generated intracellular oxidants are displayed in Figure 2A. It is known that when DCFH$_2$-DA diffuses through the cell membrane, it is de-acetylated into DCFH$_2$ inside the cytosol which is then oxidized by intracellular ROS to form fluorescent DCF [23]. As expected, treatment of HepG2 cells with AAPH significantly increased the production of intracellular ROS to 170.4% ± 3.2% relative to the NEG control. All oat bran protein hydrolysates significantly reduced ROS relative to the minus AAPH NEG control. The most significant reduction 77–104% of control normal cells was associated with pre-treatment with VPI hydrolysates, and this correlated well with their cytoprotection data (i.e. greater viability of cells). This is likely because peptides in VPI hydrolysates, or their metabolites, were transported more easily inside the cells. Similar to this work, hydrolyzed proteins from Nile tilapia were found to reduced ROS in HepG2 cells [24]; meanwhile, no effect was observed for bean globulin hydrolysates on Caco-2 cells [25]. The effect hydrolyzed proteins on intracellular ROS can then vary based on the source of proteins but also on the cellular model.

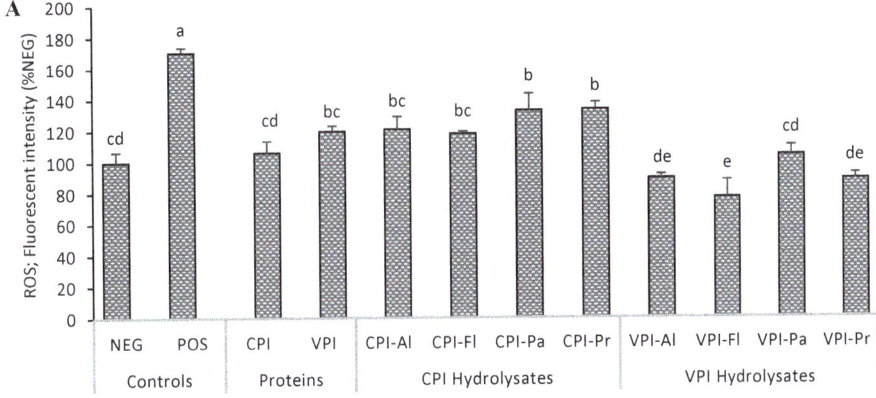

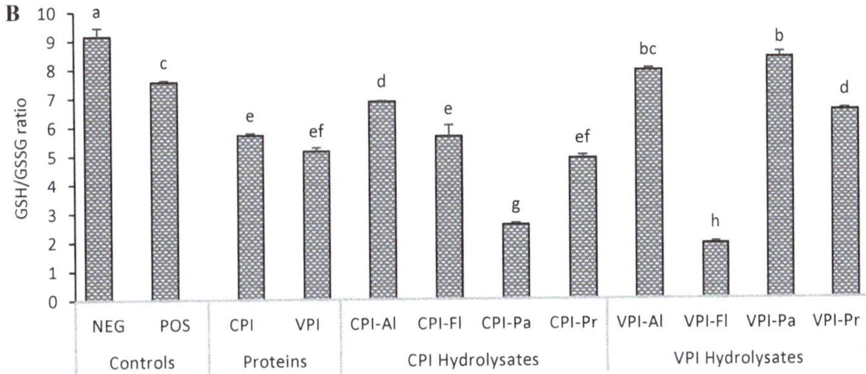

Figure 2. Intracellular reactive oxygen species (ROS) (**A**) and GSH/GSSG ratio (**B**). HepG2 cells were treated with oat bran proteins extracted with Cellulase (CPI), Viscozyme (VPI) and their Alcalase (Al), Flavourzyme (Fl), Papain (Pa) or Protamex (Pr) hydrolysates (100 µg/mL) for 24 h, followed by exposure to AAPH for 24 h, before obtaining the data. NEG: normal cells; POS: cells treated with AAPH alone. Data with different letters showed significant difference ($p < 0.05$) from Duncan's multiple range test ($n = 3$).

The oxidative status of the HepG2 cells was further evaluated by measuring the concentration of glutathione, an important cellular redox molecule. In the presence of ROS, GSH is oxidized to GSSG which can be converted back to its reduced form by glutathione reductase [26]. The ratio of GSH to GSSG is then a reliable representation of the cellular redox status. A greater ratio corresponds to less oxidative stress. In this work, HepG2 cells stressed with AAPH had a 24.0% lower ratio GSH/GSSG ratio compared to normal NEG cells (Figure 2B), indicating partial depletion of reduced GSH which might translate into an inability to maintain the redox balance. However, when HepG2 cells were pre-treated with hydrolysates, VPI-Al and VPI-Pa resulted in a higher ratio of GSH/GSSG than the AAPH NEG control group (without AAPH). In the presence of other hydrolysates, the cellular environment was more oxidized as GSH/GSSG ratios were similar or lower compared to AAPH control. Cells pre-treated with VPI-Al and VPI-Pa, and which have a higher ratio of GSH/GSSH, are amongst those showed less intracellular ROS and greater viability. This likely gives these cells the greatest ability to maintain redox balance.

In our previous work, tested the radical scavenging activities of hydrolysates used in the present study and determined the sequences of peptides in CPI-Pa, VPI-Pa, and VPI-Pr using tandem mass spectrometry [13]. It was found that CPI-Pa and VPI-Pa had similar scavenging power for peroxyl radicals; meanwhile, VPI-Pa had higher superoxide anion radical scavenging and ferrous iron chelating activities. The reason was attributed to higher ratios of charged residues and histidine in VPI-Pa relative to CPI-Pa but also its higher content of peptides with less than 10 residues (42% in VPI-Pa and 37% in CPI-Pa). Smaller peptides in VPI-Pa might then explained its better protection in most assays (e.g., ROS, GSH, cytoprotection) against AAPH-induced cellular injury. This is likely because smaller peptides are less degraded by cellular proteases.

3.3. Antioxidant Enzymes Activity

The enzymes catalase, glutathione peroxidase (GPx) and superoxide dismutase (SOD) are important in the prevention of oxidative stress through their action on ROS which are converted into stable or less reactive species. The hydrolysates, CPI-Fl, VPI-Al, VPI-Fl, VPI-Pa, and VPI-Pr were selected for this section because they were either cytoprotective, produced less ROS, or had a higher amount of GSH. The induction of oxidative stress by AAPH reduced the activity of catalase, glutathione peroxidase (GPx), and superoxide dismutase (SOD) to 87.5% ± 6.1%, 75.7% ± 0.7% and 55.2% ± 1.1%, respectively in relation to normal cells (Figure 3). The change was, however, not significant in the case of catalase. The activity of catalase increased by up to 3-fold when cells were pre-treated either of the five hydrolysates with VPI-Pa having the most up-regulation (Figure 3A). A similar increase was reported in HepG2 due to the action of peptide GLVYIL [23].

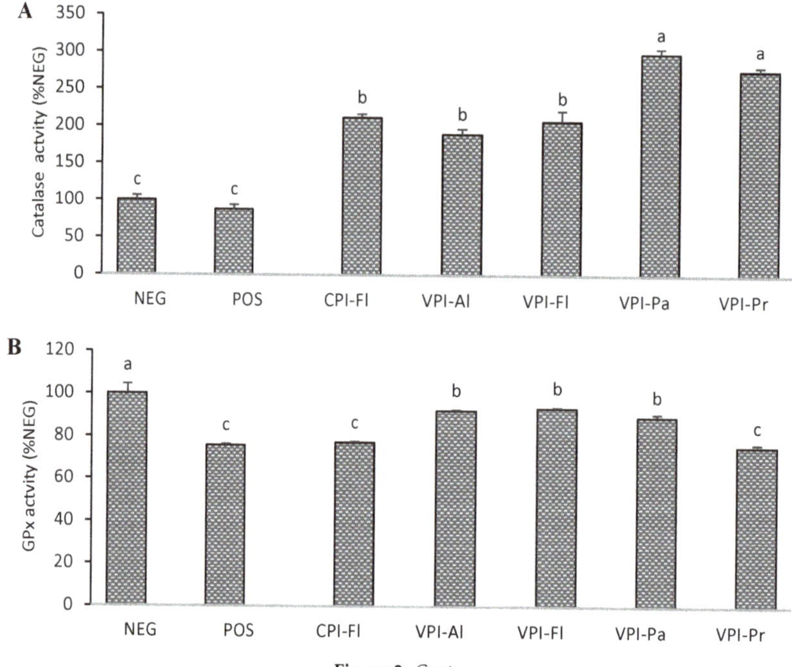

Figure 3. *Cont.*

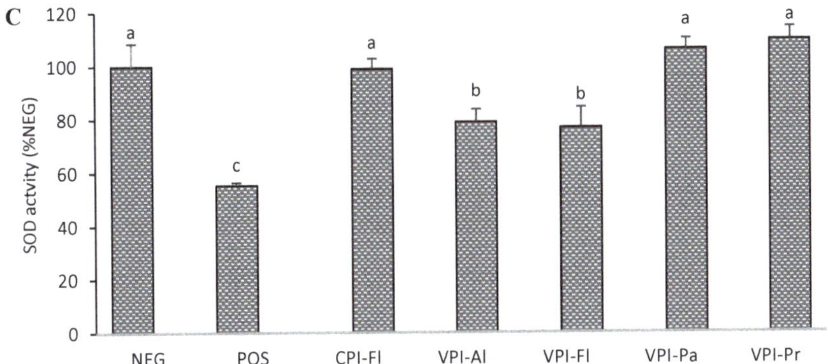

Figure 3. Activity of the antioxidant enzymes in AAPH-stressed HepG2 cells of hydrolysates (100 µg/mL) that showed lower reactive oxygen species (ROS) and higher GSH/GSSG ratios. Catalase (**A**); glutathione peroxidase (**B**); superoxide dismutase (**C**). NEG: normal cells (no treatment); POS: positive control (AAPH treated only). Oat bran proteins extracted with Cellulase (CPI) or with Viscozyme (VPI) and then hydrolyzed with Alcalase (Al), Flavourzyme (Fl), Papain (Pa) or Protamex (Pr). Data with different letters showed significant difference ($p < 0.05$) from Duncan's multiple range test ($n = 3$).

Figure 3B showed partial recovery of GPx activity in HepG2 when they were pre-treated with three (VPI-Al, VPI-Fl and VPI-Pa) of the five hydrolysates (Figure 3B). There was no change in GPx activities of CPI-Fl and VPI-Pr. GPx is important for the conversion of hydrogen peroxide and hydroperoxides, which are reactive molecules, into the water and hydroxylated molecules. The activity of GPx is also dependent on GSH supply and is therefore not surprising that the hydrolysates with higher activities are also the ones with greater GSH contents. Other studies have also reported a positive effect of food protein hydrolysates (e.g. rice, fish) on the activity of GPx in hepatocyte cells [27,28].

In the SOD assay, three of the hydrolysates (CPI-FL, VPI-Pa, and VPI-Pr) completely prevented the adverse effects of AAPH-induced oxidative stress of the enzyme activity (Figure 3C). The two others (VPI-Fl, VPI-AL) partially prevented the loss of SOD activities but, this was still significant. Casein hydrolysates had an enhancing effect on SOD activities in lymphocyte cells [29] while a similar effect was observed for fish gelatin hydrolysates in human colon cells [22].

Catalase is more resistant to inactivation by peroxyl radicals due to the narrowness of its active site which can explain why it was the least affected by AAPH. It has however been reported that lipid peroxides might cause induction of the catalase [30] which explained the 3-fold increase found in this work. SOD converts superoxide anion radical to hydrogen peroxide which is subsequently converted to water by catalase. The two hydrolysates that produced the highest SOD activities also led to highest catalase activities, and this is a relevant presumption as proper elimination of hydrogen peroxide is important to prevent the loss of SOD activity. It is possible that peptides present in oat protein hydrolysates may have induced the expression of catalase and SOD genes as reported in a recent work [31].

The difference in composition of peptides and their size might explain their behaviour on the cellular antioxidant enzymes. In the previous work, larger peptides in VPI-Pr contained 26 residues compared to 20 in VPI-Pa while 31% of peptides in VPI-Pr were less than 10 residues relative to 42% in VPI-Pa which was due to higher proteolysis activity of Papain [13]. The enzymes tested in this work are metalloenzymes, peptides in hydrolysates that interfere in metal (iron, selenium, zinc or manganese) update or metabolism can then affect the activity of catalase, GPx or SOD. For example, it has been reported that iron uptake by cells was stimulated by radicals [32] and this could explain the increased of catalase activity we observed in the presence of AAPH. It was also found that the decrease in GPx activity in the liver was associated in impairment of selenium absorption [33]. Peptides in hydrolyzed oat proteins might have prevented the oxidation of the selenocysteine at the active site of GPx.

3.4. Caspase-3 Determination and Cell Apoptosis

Caspase-3 enzyme belongs to a family of proteases that exists in an inactive form, and its activation is highly regulated. These proteases are responsible for programmed cell death, control cell disintegration and prevent the release of their content into the intracellular matrix which will damage the surrounding cells [34]. Caspase-3, one of the executioner caspases, is important specifically since its presence is mostly an indicator of irreversible cell death [34]. This enzyme can cleave Asp-Glu-Val-Asp *p*-nitroanilide (DEVD-pNA) and release *p*-nitroaniline (pNA). The absorbance of pNA is proportional to the enzyme activity, and hence, higher intensities indicate more apoptosis.

Treatment of HepG2 cells with AAPH increased the activity of caspase-3 by approximately 3-fold (Figure 4), which translated in greater proteolytic activity and therefore greater cells death. Pre-treatment of cells with the hydrolysates resulted in a reduction of 2- to 3-fold in caspase-3 activity, relative to the AAPH NEG control. Only one the hydrolysates, CPI-Fl, restored caspase-3 activity to the level found in the non-treated group. There are different mechanisms proposed as for how the antioxidant protein hydrolysates can exert their effect. Some of the hydrolysates with higher cell viability and greater catalase and SOD activities did not possess lower caspase-3 activities, likely because there are different anti-apoptotic mechanisms. Whey protein hydrolysates, for example, showed an anti-apoptotic effect in PC12 cells by up-regulating the expression of anti-apoptotic Bcl-2, a regulatory protein that can proteolytically cleave pro-apoptotic proteins and reduce the expression of the pro-apoptotic Bax protein [35]. In endothelial cells, the anti-apoptotic properties of seahorse protein hydrolysates down-regulated caspase-3 and p53 and increased the Bcl-2/Bax ratio [36]. Another potential mechanism is the upregulation of poly (ADP-ribose) polymerase which is responsible for DNA repair [37]. In addition to caspase-3, some of the above mechanisms might have been involved in the protection provided by the hydrolyzed oat bran proteins. Peptides in hydrolyzed oat proteins might then have protected HepG2 cells through some of these mechanisms rather than through caspase-3 inhibition.

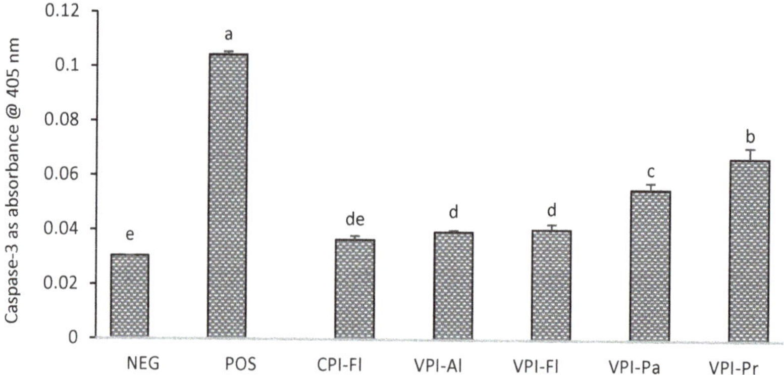

Figure 4. Changes in absorbance corresponding to the change in the activity of caspase-3. HepG2 were hydrolysates (100 µg/mL) for 24 followed by exposure to AAPH for 24 h before determining the activity using a Kit. Oat bran proteins extracted with Cellulase (CPI) or with Viscozyme (VPI) and then hydrolyzed with Alcalase (Al), Flavourzyme (Fl), Papain (Pa) or Protamex (Pr). NEG: normal cells (no treatment); POS: positive control (treated with AAPH only). Data with different letters showed significant difference ($p < 0.05$) from Duncan's multiple range test ($n = 3$).

4. Conclusions

Oat bran protein isolates, and their enzymatic digests, protected hepatic HepG2 cells from induced oxidative stress. The protection of protein isolated was dependent on the extraction procedure, as well as whether Cellulase or Viscozyme was used. The protection of hydrolysates was related to the extraction procedure and the protease utilized. Hydrolysates derived from Viscozyme extracted proteins had the highest antioxidant activities; specifically, those of Papain and Flavouzyme. Future work will investigate the activity of peptides present in these hydrolysates.

Author Contributions: Conceptualization, R.E., W.G.W. and A.T.; methodology, R.E., W.G.W. and A.T.; software, R.E.; validation, R.E. and A.T.; formal analysis, R.E.; writing—original draft preparation, R.E.; writing—review and editing, W.G.W. and A.T.; supervision, W.G.W. and A.T.

Funding: This research was funded by Discovery Grants from the National Science and Engineering Research Council of Canada (NSERC) to A.T. (No.: 371908) and W.G.W.

Conflicts of Interest: The authors declare no conflict of interest.

References

1. Cichoz-Lach, H.; Michalak, A. Oxidative stress as a crucial factor in liver diseases. *World J. Gastroenterol.* **2014**, *20*, 8082–8091. [CrossRef]
2. Leung, R.; Venus, C.; Zeng, T.; Tsopmo, A. Structure-function relationships of hydroxyl radical scavenging and chromium-VI reducing cysteine-tripeptides derived from rye secalin. *Food Chem.* **2018**, *254*, 165–169. [CrossRef] [PubMed]
3. Moronta, J.; Smaldini, P.L.; Docena, G.H.; Añón, M.C. Peptides of amaranth were targeted as containing sequences with potential anti-inflammatory properties. *J. Funct. Foods* **2016**, *21*, 463–473. [CrossRef]
4. Yang, T.; Zhu, H.; Zhou, H.; Lin, Q.; Li, W.; Liu, J. Rice protein hydrolysate attenuates hydrogen peroxide induced apoptosis of myocardiocytes H9c2 through the Bcl-2/Bax pathway. *Food Res. Int.* **2012**, *48*, 736–741. [CrossRef]
5. Li, S.; Tan, H.Y.; Wang, N.; Cheung, F.; Hong, M.; Feng, Y. The potential and action mechanism of polyphenols in the treatment of liver diseases. *Oxid. Med. Cell. Longev.* **2018**, *2018*, 8394818. [CrossRef]
6. Walters, M.E.; Esfandi, R.; Tsopmo, A. Potential of food hydrolyzed proteins and peptides to chelate iron or calcium and enhance their absorption. *Foods* **2018**, *7*, 172. [CrossRef] [PubMed]
7. Yu, G.-C.; Lv, J.; He, H.; Huang, W.; Han, Y. Hepatoprotective effects of corn peptides against carbon tetrachloride-induced liver injury in mice. *J. Food Biochem.* **2012**, *36*, 458–464. [CrossRef]
8. Moritani, C.; Kawakami, K.; Fujita, A.; Kawakami, K.; Hatanaka, T.; Tsuboi, S. Anti-oxidative activity of hydrolysate from rice bran protein in HepG2 cells. *Biol. Pharm. Bull.* **2017**, *40*, 984–991. [CrossRef] [PubMed]
9. Kang, K.-H.; Qian, Z.-J.; Ryu, B.; Karadeniz, F.; Kim, D.; Kim, S.-K. Antioxidant peptides from protein hydrolysate of microalgae Navicula incerta and their protective effects in Hepg2/CYP2E1 cells induced by ethanol. *Phytother. Res.* **2012**, *26*, 1555–1563. [CrossRef] [PubMed]
10. Tsopmo, A.; Cooper, A.; Jodayree, S. Enzymatic hydrolysis of oat flour protein isolates to enhance antioxidative properties. *Adv. J. Food Sci. Technol.* **2010**, *2*, 206–212.
11. Vanvi, A.; Tsopmo, A. Pepsin digested oat bran proteins: separation, antioxidant activity, and identification of new peptides. *J. Chem.* **2016**, *2016*, 8216378. [CrossRef]
12. Baakdah, M.M.; Tsopmo, A. Identification of peptides, metal binding and lipid peroxidation activities of HPLC fractions of hydrolyzed oat bran proteins. *J. Food Sci. Technol.* **2016**, *53*, 3593–3601. [CrossRef]
13. Esfandi, R.; Willmore, W.G.; Tsopmo, A. Peptidomic analysis of hydrolyzed oat bran proteins, and their in vitro antioxidant and metal chelating properties. *Food Chem.* **2019**, *279*, 49–57. [CrossRef]
14. Liu, Y.; Nair, M.G. An efficient and economical MTT assay for determining the antioxidant activity of plant. *J. Nat. Prod.* **2010**, *73*, 1193–1195. [CrossRef] [PubMed]
15. Vieira, E.F.; da Silva, D.D.; Carmo, H.; Ferreira, I.M.P.L.V.O. Protective ability against oxidative stress of brewers' spent grain protein hydrolysates. *Food Chem.* **2017**, *228*, 602–609. [CrossRef]
16. Wolfe, K.L.; Rui, H.L. Cellular antioxidant activity (CAA) assay for assessing antioxidants, foods, and dietary supplements. *J. Agric. Food Chem.* **2007**, *55*, 8896–8907. [CrossRef]
17. Rahman, I.; Kode, A.; Biswas, S.K. Assay for quantitative determination of glutathione and glutathione disulfide levels using enzymatic recycling method. *Nat. Protoc.* **2006**, *1*, 3159–3165. [CrossRef]
18. Beers, R.F.; Sizer, I.W. A spectrophotometric method for measuring the breakdown of hydrogen peroxide by catalase. *J. Biol. Chem.* **1952**, *195*, 133–140. [PubMed]

19. Lawrence, R.A.; Burk, R.F. Glutathione peroxidase activity in selenium-deficient rat liver. *Biochem. Biophys. Res. Commun.* **1976**, *71*, 952–958. [CrossRef]

20. Spitz, D.R.; Oberley, L.W. An assay for superoxide dismutase activity in mammalian tissue homogenates. *Anal. Biochem.* **1989**, *179*, 8–18. [CrossRef]

21. Franek, F.; Hohenwarter, O.; Katinger, H. Plant protein hydrolysates: Preparation of defined peptide fractions promoting growth and production in animal cells cultures. *Biotechnol. Prog.* **2000**, *16*, 688–692. [CrossRef] [PubMed]

22. Karnjanapratum, S.; O'Callaghan, Y.C.; Benjakul, S.; O'Brien, N. Antioxidant, immunomodulatory and antiproliferative effects of gelatin hydrolysate from unicorn leatherjacket skin. *J. Sci. Food Agric.* **2016**, *96*, 3220–3226. [CrossRef] [PubMed]

23. Du, Y.; Esfandi, R.; Willmore, W.G.; Tsopmo, A. Antioxidant activity of oat proteins derived peptides in stressed hepatic hepg2 cells. *Antioxidants* **2016**, *5*, 39. [CrossRef]

24. Yarnpakdee, S.; Benjakul, S.; Kristinsson, H.G.; Bakken, H.E. Preventive effect of Nile tilapia hydrolysate against oxidative damage of HepG2 cells and DNA mediated by H_2O_2 and AAPH. *J. Food Sci. Technol.* **2015**, *52*, 6194–6205. [CrossRef] [PubMed]

25. Carrasco-Castilla, J.; Hernández-Álvarez, A.J.; Jiménez-Martínez, C.; Jacinto-Hernández, C.; Alaiz, M.; Girón-Calle, J.; Dávila-Ortiz, G. Antioxidant and metal chelating activities of Phaseolus vulgaris L. var. Jamapa protein isolates, phaseolin and lectin hydrolysates. *Food Chem.* **2012**, *131*, 1157–1164. [CrossRef]

26. Forman, H.J.; Zhang, H.; Rinna, A. Glutathione: Overview of its protective roles, measurement, and biosynthesis. *Mol. Aspects Med.* **2009**, *30*, 1–12. [CrossRef]

27. Tao, J.; Zhao, Y.Q.; Chi, C.F.; Wang, B. Bioactive Peptides from Cartilage Protein Hydrolysate of Spotless Smoothhound and Their Antioxidant Activity In Vitro. *Marine Drugs* **2018**, *16*, 100. [CrossRef]

28. Han, B.K.; Park, Y.; Choi, H.S.; Suh, H.J. Hepatoprotective effects of soluble rice protein in primary hepatocytes and in mice. *J. Sci. Food Agric.* **2016**, *96*, 685–694. [CrossRef]

29. Phelan, M.; Aherne-Bruce, S.A.; O'Sullivan, D.; FitzGerald, R.J.; O'Brien, N.M. Potential bioactive effects of casein hydrolysates on human cultured cells. *Int. Dairy J.* **2009**, *19*, 279–285. [CrossRef]

30. Meilhac, O.; Zhou, M.; Santanam, N.; Parthasarathy, S. Lipid peroxides induce expression of catalase in cultured vascular cells. *J. Lipid Res.* **2000**, *41*, 1205–1213.

31. Homayouni-Tabrizi, M.; Asoodeh, A.; Soltani, M. Cytotoxic and antioxidant capacity of camel milk peptides: Effects of isolated peptide on superoxide dismutase and catalase gene expression. *J. Food Drug Anal.* **2017**, *25*, 567–575. [CrossRef]

32. Richardson, D.R.; Ponka, P. Identification of a mechanism of iron uptake by cells which is stimulated by hydroxyl radicals generated via the iron-catalysed Haber-Weiss reaction. *Biochim. Biophys. Acta* **1995**, *1269*, 105–114. [CrossRef]

33. Bourre, J.-M.; Dumont, O.; Clément, M.; Dinh, L.; Droy-Lefaix, M.-T.; Christen, Y. Vitamin E deficiency has different effects on brain and liver phospholipid hydroperoxide glutathione peroxidase activities in the rat. *Neurosci. Lett.* **2000**, *286*, 87–90. [CrossRef]

34. Harrington, H.A.; Ho, K.L.; Ghosh, S.; Tung, K. Construction and analysis of a modular model of caspase activation in apoptosis. *Theor. Biol. Med. Model.* **2008**, *5*, 1–15. [CrossRef] [PubMed]

35. Jin, M.M.; Zhang, L.; Yu, H.X.; Meng, J.; Sun, Z.; Lu, R.R. Protective effect of whey protein hydrolysates on H_2O_2-induced PC12 cells oxidative stress via a mitochondria-mediated pathway. *Food Chem.* **2013**, *141*, 847–852. [CrossRef] [PubMed]

36. Oh, Y.; Ahn, C.B.; Yoon, N.Y.; Nam, K.H.; Kim, Y.K.; Je, J.Y. Protective effect of enzymatic hydrolysates from seahorse (Hippocampus abdominalis) against H_2O_2-mediated human umbilical vein endothelial cell injury. *Biomed. Pharmacother.* **2018**, *108*, 103–110. [CrossRef] [PubMed]

37. Redza-Dutordoir, M.; Averill-Bates, D.A. Activation of apoptosis signalling pathways by reactive oxygen species. *Biochim. Biophys.* **2016**, *1863*, 2977–2992. [CrossRef]

Review

Extraction and Characterization of Antioxidant Peptides from Fruit Residues

Saúl Olivares-Galván, María Luisa Marina and María Concepción García *

Departamento de Química Analítica, Química Física e Ingeniería Química, Instituto de Investigación Química "Andrés M. del Río", Universidad de Alcalá, Ctra. Madrid-Barcelona, km. 33.600, E-28871 Alcalá de Henares, Madrid, Spain; saul.olivares@uah.es (S.O.-G.); mluisa.marina@uah.es (M.L.M.)
* Correspondence: concepcion.garcia@uah.es

Received: 5 June 2020; Accepted: 21 July 2020; Published: 29 July 2020

Abstract: Fruit residues with high protein contents are generated during the processing of some fruits. These sustainable sources of proteins are usually discarded and, in all cases, underused. In addition to proteins, these residues can also be sources of peptides with protective effects against oxidative damage. The revalorization of these residues, as sources of antioxidant peptides, requires the development of suitable methodologies for their extraction and the application of analytical techniques for their characterization. The exploitation of these residues involves two main steps: the extraction and purification of proteins and their hydrolysis to release peptides. The extraction of proteins is mainly carried out under alkaline conditions and, in some cases, denaturing reagents are also employed to improve protein solubilization. Alternatively, more sustainable strategies based on the use of high-intensity focused ultrasounds, microwaves, pressurized liquids, electric fields, or discharges, as well as deep eutectic solvents, are being implemented for the extraction of proteins. The scarce selectivity of these extraction methods usually makes the subsequent purification of proteins necessary. The purification of proteins based on their precipitation or the use of ultrafiltration has been the usual procedure, but new strategies based on nanomaterials are also being explored. The release of potential antioxidant peptides from proteins is the next step. Microbial fermentation and, especially, digestion with enzymes such as Alcalase, thermolysin, or flavourzyme have been the most common. Released peptides are next characterized by the evaluation of their antioxidant properties and the application of proteomic tools to identify their sequences.

Keywords: fruit residues; antioxidant; peptides; extraction

---

1. Introduction

The growing world population, together with the increasing popular awareness about healthy nutritional habits, has promoted a massive rise in fruit production [1–3]. This trend has boosted the release of fruit residues. A production residue is defined as a material that is not deliberately produced in a productive process. If that residue has a certain use, is ready for use without further processing, and has to be produced as an integral part of the production process, then that residue is called "by-product." If any of those three conditions are not met, the residue is called "waste" [4].

It has been estimated that approximately 50% of the original weight of fruits becomes waste in the form of peels, pomaces, seeds, and unripe or damaged fruits, which is an unsustainable rate [5]. The usual strategies for the management of these wastes include landfilling and incineration. These practices demand high amounts of oxygen, are associated with the emission of greenhouse gases, and create a platform for pathogenic bacteria and pests. Furthermore, unacceptable odors are generated during their biodegradation [6,7]. Nevertheless, many of these wastes can be reused, and some are currently employed in composting and animal feeding. A more efficient use for these resources is as feedstock in biorefineries, replacing petrochemical-based matter to produce high value-added products such as chemicals, materials, and fuels [8].

Additionally, many research works have shown that fruit byproducts contain large amounts of many phytochemicals and essential nutrients. Pectin, polyphenols, carotenoids, flavonoids, and fiber are some of the functional and nutritional constituents of fruit residues that have attracted the greatest interest [9]. However, bioactive proteins and peptides, which have been more explored in foods from animal origin, are also present in plants [10]. Fruit seeds, for example, are usually the main constituents of fruit residues and store large quantities of proteins and peptides, as well as lipids and carbohydrates, which constitute the plant's food reserves in its first stages of growth [11].

2. Antioxidant Peptides in Fruit Residues

Bioactive peptides can be defined as food components with a positive effect on body functions or conditions beyond their nutritional effects and that may ultimately influence health in a positive manner [12]. A bioactive peptide usually contains 2–20 amino acid residues and can exhibit different biological functions depending on its chemical structure, length, and amino acid composition [13]. Regardless of the kind of sample, peptides can exist as an independent entity or, more often, in a latent state as part of the protein sequence. In this case, the release of peptides requires the hydrolysis of parent proteins [12,14].

Bioactive peptides can be added to foods to improve their functionality (functional foods) or can be used in the manufacture of nutraceuticals. Peptides have shown different bioactivities, including antimicrobial, anticancer, antiviral, hemolytic, and antihypertensive activity, among others, with antioxidant activity being one of the most researched [15].

Oxidative stress is caused by the presence of high amounts of reactive oxygen species (ROS) that overcome endogenous antioxidant defense mechanisms. Maintained oxidative stress can lead to the development of serious diseases by damaging important biomolecules such as DNA, proteins, and lipids [16]. Different studies have demonstrated that antioxidant intake is inversely related to cellular death, aging, and the development of diseases such as diabetes and cancer [17].

In addition to the health benefits, antioxidant peptides can be added to food systems to reduce oxidative changes during storage [18]. While lipid peroxidation inhibition is the most important mechanism, peptides are also capable of reducing the oxidative modification of intact proteins. Vegetable protein hydrolysates are already allowed to be used as food additives in the United States. Moreover, antioxidant peptides could also be added to cosmeceutical products to neutralize free radicals, thus preventing the signs of aging skin [19].

There are some common features within antioxidant peptides. They usually present a large amount of hydrophobic amino acids, such as leucine, alanine, and phenylalanine, that enhance hydrogen-transfer and lipid peroxyl radical trapping, promote their accessibility to hydrophobic targets, and make it easier to pass through cell membranes [12,16,20]. On the other hand, the presence of aromatic amino acids, such as histidine, tyrosine, and tryptophan within a peptide sequence has also been found to be related to antioxidant properties due to their capability to donate electrons to free radicals, thus converting them into stable molecules [20,21]. However, aromatic and hydrophobic amino acids impart a bitter taste to protein hydrolysates, which may create organoleptic problems when used as food additives [18].

Molecular weights between 0.5 and 1.5 kDa are also a common feature within antioxidant peptides [22]. Antioxidant peptides have been obtained from rapeseed residues (*Brassica napus*) [23,24], peels of pomegranate (*Punica granatum*) [3,25] and mango (*Mangifera indica*) [26], and seeds of apricot (*Prunus armeniaca*) [11,27], peach (*Prunus persica L.*) [11,16,28,29], bottle gourd (*Lagenaria sciceraria*) [30–32], cherry (*Prunus cerasus L.*) [11,33], olive (*Olea europaea*) [11,16,17,34], plum (*Prunus domestica L.*) [11,14,29,35], tomato (*Solanum lycopersicum*) [21,36–43], wax gourd (*Benincasa hispida*) [44], jujube (*Ziziphus jujube*) [45,46], muskmelon (*Cucumis melo*) [47], watermelon (*Citrullus lanatus*) [32,48–53], papaya (*Carica papaya*) [54], Chinese cherry (*Prunus pseudocerasus*) [55], African breadfruit (*Treculia*

Africana) [56], pumpkin (*Cuburbita pepo* [57–62] and *Cucurbita moschata* [20,32,63,64]), and date (*Phoenix dactylifera*) [65,66]. Some of these residues are released in the extraction of the oil fraction of seeds (such as pumpkin seeds or rapeseed) and of fruits (such as olives) or during the processing of fruits or vegetables such as pomegranate, *Prunus* fruits, tomato, muskmelon, watermelon, papaya, and mango.

3. Obtaining Antioxidant Proteins and Peptides from Fruit Residues

The exploitation of sustainable sources of proteins requires the development of suitable methodologies. A general procedure followed for obtaining proteins and peptides with antioxidant properties from fruit residues is shown in Figure 1. Usually, the fruit residue is dried and ground before the extraction in order to avoid microbiological contamination during storage and to promote the penetration of the extracting solvent into the solid matrix. Furthermore, when the lipid content in the fruit residue is high and disturbs the extraction of proteins, it is necessary to include a previous defatting step. Typically, a preliminary extraction with hexane is the selected procedure. Sometimes, there is also a sieving step before the extraction to obtain a more homogeneous material [21,43,45–47]. Then, the extraction of proteins is performed.

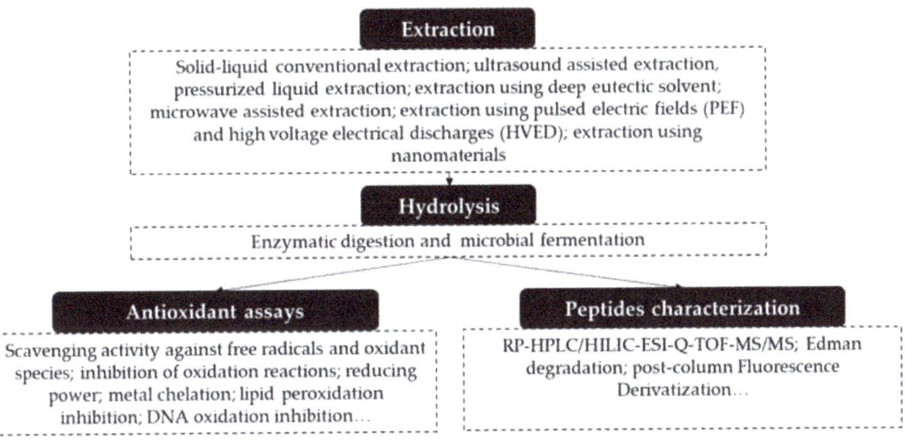

Figure 1. General procedure employed for obtaining proteins and peptides from fruit residues. Rp-HPLC: reversed-phase HPLC; HILIC: hydrophilic interaction chromatography; ESI: electrospray ionization; and Q-TOF: quadrupole time-of-flight.

Conventional and non-conventional techniques have been tested for the extraction of proteins from fruit residues. Protein extraction from plant tissues is currently carried out by strategies that involve the use of polluting reagents and volatile organic solvents, and they result in very low yields. Moreover, some of these reagents are not food-grade and cannot be employed in industrial applications. Protein extraction from sustainable sources, such as fruit residues, urges the development of alternative strategies with a lower environmental impact and a higher protein yield. The application of sustainable techniques that require less polluting reagents and less energy are of special interest. Another important aspect in protein extraction is selectivity. The usual lack of the selectivity of extraction procedures makes, in many cases, an additional step to purify proteins necessary.

Extracted proteins are usually submitted to a hydrolysis process to obtain peptides. This step generally requires the use of food grade enzymes. Extracted proteins and peptides are sometimes fractionated based on different parameters—mainly molecular mass, solubility, and hydrophobicity—to have a deeper inside on properties of proteins/peptides within different fractions.

The evaluation of antioxidant properties in extracts, hydrolysates, and/or fractions involves the use of different in vitro assays based on different mechanisms. Additional studies evaluating the capacity of proteins and peptides to reduce the oxidative stress on cells cultures and animal models are very interesting to confirm in vitro results. In some cases, a further characterization of extracts is carried out by the identification of peptides using tandem mass spectrometry.

4. Techniques Used in the Extraction and Purification of Proteins

Protein extraction requires the breakdown of tissues, cell membranes, and cell walls in order to release intracellular material. The difficulty is high in the case of plant tissues due to the presence of large vacuoles, the rigidity and thickness of cell walls, and the heterogeneity of proteins. Moreover, the presence of lipids, polysaccharides, or phenolic compounds can interfere with the extraction of proteins [14].

The amount and characteristics of proteins in fruit residues is highly variable being not possible to generalize. Indeed, the dried peels of pawpaw, pineapple, mango, apple, banana, orange, pomegranate, and watermelon present between 2.8%, for apples, and 18.1%, for pawpaw, of crude proteins [67]. On the other hand, fruit seeds, in general, show a higher protein content. For example, the seeds of cherry, pumpkin, papaya, watermelon, mango, jackfruit, orange, melon, peach, and Surinam cherry present a protein content ranging from 6%, for the mango seeds, to 39%, for the pumpkin seeds [68,69]. Some functional proteins have been found in some residues, such as passiflin, a dimeric protein from passion fruit seeds that exhibits antifungal and anticancer activities [70].

The performance and sustainability of protein extraction can be improved by favoring physical contact between the extracting medium and proteins and by using more environmentally friendly solvents. Physical contact between the extracting medium and target compounds can be promoted by the use of ultrasound-assisted extraction, pressurized liquid extraction, microwave-assisted extraction, or by the application of electric energy (pulsed electric fields or high-voltage electrical discharges). These techniques are being implemented in the extraction of proteins from fruit residues at the laboratory scale. Additionally, the introduction of nanomaterials in the extraction and purification of proteins and the use of deep eutectic solvents are promising approaches to increase the sustainability of extraction procedures. Table 1 summarizes the methods employed to extract proteins and peptides from fruit residues.

4.1. Solid–Liquid Conventional Extraction

Traditional methods for the extraction of proteins employ aqueous buffers that can contain reducing and chaotropic reagents (dithiothreitol, mercaptoethanol, urea, etc.), surfactants, etc. Moreover, protein extraction is usually followed by a final purification step that involves the use of an organic solvent or the acidification of the sample.

Since most food proteins have low isoelectric points, the extraction of proteins at pHs ranging from 7.5 to 12 using NaOH solutions and stirring, followed by their acidic precipitation at pHs from 3.8 to 5.3 is very popular. Incubation temperatures of 40–50 °C have been employed to promote protein solubilization. This strategy has been used in the extraction of proteins from the seeds of tomato, jujube, watermelon, Chinese cherry, African breadfruit, and pumpkin, as well as in pumpkin oil cake [21,37,41–43,45,48,50,52,55–62]. Despite the popularity and industrial applicability of this procedure, its selectivity and extraction yield, in general, are very low. Extraction yield has been found to range from 23%, in the case of the African breadfruit seeds, to 56%, in the case of the pumpkin seeds [56,59]. After a purification step by acid precipitation, the protein contents of isolates reached 80%, 82%, and 90% for tomato, Chinese cherry, and African breadfruit seeds, respectively [42,55,56]. In addition to the acid precipitation of proteins, other purification protocols using acetone or $(NH_4)_2SO_4$ have been employed in the case of the jujube seeds [45,46]. The main advantage of using $(NH_4)_2SO_4$ is its non-denaturing character. Additionally, alkaline extraction results in protein degradation, the reduction of protein solubility at neutral pH, and poor technological functionalities.

Alternatively, proteins have been extracted at other pHs. Parniakov et al. [26,54] used different pHs (2.5, 6.0, and 11.0) for extracting proteins from mango peels and papaya seeds, but, again, they observed the highest extraction at pH 11.0. Phosphate and Tris(hydroxymethyl)aminomethane (Tris)-HCl buffers were used in the extraction of milled rapeseed, bottle gourd, pumpkin (*Cucurbita moschata*), watermelon, wax gourd, and jujube seeds [23,30,32,44,45]. Under the same extraction conditions, pumpkin (*Cucurbita moschata*), watermelon, and bottle gourd seeds yielded isolates with 46%, 39%, and 49% protein contents, respectively. In some cases, buffers contained additives to avoid protease activity (ethylenediaminetetraacetic acid; EDTA) or to promote protein solubilization. Indeed, proteins are folded and usually form insoluble aggregates that constitute a limitation for their extraction. Dithiothreitol (DTT) is a usual additive that reduces disulfide bonds between cysteine residues and, thus, improves the extraction of proteins. Urea, for example, is a chaotropic agent that is added to disrupt hydrogen bonds between amino acids. Surfactants, such as SDS and Triton X-100 have also been added to the extracting media at low concentrations (below the critical micelle concentration). SDS is a denaturing surfactant that disrupts cell membranes and breaks interactions within proteins. Triton X-100 is a non-denaturing surfactant that cannot penetrate into proteins and disrupt interactions, but it can associate with hydrophobic parts of the protein to promote solubilization [23,30,44]. While extractions in an alkalized medium is an usual procedure in the food industrial environment, the use of a Tris-HCl buffer is not suitable. Moreover, additives such us DTT, SDS, and urea are not food-grade reagents and cannot be used in the manufacture of products for animal or human consumption.

The Osborne method [71] has also been employed for the extraction and fractionation of proteins based on their solubility in different media: water (albumins), salt solution (globulins), alkaline solution (glutelins), and alcoholic solution (prolamins). This methodology was applied for the extraction of proteins from pumpkin (*Cucurbita moschata)*, bottle gourd, muskmelon, and watermelon seeds [20,31,47,49]. In all cases, seeds were previously defatted obtaining powders with 56–69% proteins. Most proteins were extracted using salt or alkaline solutions, while lower proteins were extracted using alcoholic solutions. Pumpkin seeds, for example, resulted in a globulin fraction that held the highest protein content (46%), followed by the glutelin fraction (39%), the albumin fraction (23%), and the prolamin fraction (12%). The trend of showing a high protein globulin fraction and a low protein prolamin fraction was common with other seeds such as the watermelon seeds and bottle gourd seeds. Unlike them, melon seeds showed the highest protein content in the glutelin fraction (81%), although the prolamin (6%) fraction was, again, the fraction with the lowest protein content.

4.2. Ultrasound-Assisted Extraction of Proteins

Extraction using high-intensity focused ultrasound (HIFU), first developed around 1950, has been widely employed for the acceleration of these procedures. HIFU provides mechanical energy in the form of acoustic energy and extraction is based on a phenomenon known as cavitation. Ultrasonic waves generate rapid changes in pressure within a solution that lead to the formation of small gas bubbles that collapse and thereby release a high amount of energy. This energy promotes the breakdown of tissues and cell walls, followed by the extraction of proteins [72].

The HIFU extraction of proteins has often been carried out with a Tris-HCl buffer that could also contain SDS and DTT at low concentrations (0.5–1% and 0.1–0.5%, respectively). The optimization of this process required tuning the extraction time, the ultrasound amplitude, the concentration of SDS and DTT, and the sample:solvent ratio. Under optimal conditions, HIFU has been employed for the extraction of proteins from *Prunus* fruits (plum, peach, cherry, and apricot), olive seeds [11,14,16,17,29,33,35], and pomegranate peels [3]. An ultrasound amplitude of 30% has enabled the reduction of extraction times from hours to 1–5 min.

However, despite being an old technique, upscaling to pilot or industrial use has failed to succeed to date [73], and the solvent and additives used here are not suitable for food production.

Table 1. Methods employed in the extraction and purification of proteins from fruit residues.

Residue	Extracting Media	Extraction Conditions	Protein Purification Conditions	Ref.
SOLID–LIQUID CONVENTIONAL EXTRACTION				
Tomato seeds	NaCl (1.5%, pH 11.5)	DGS: extracting medium at ratio 1:10 (w/v); stirring (room temperature and 1 h)	Centrifugation and precipitation at pH 4.0	[21,43]
Tomato seeds	NaOH aq. (pH 7.5–1.5)	1 g of DGS and 82 mL water; stirring (50 °C and 50 h)	Centrifugation and filtration	[37,41]
Tomato seeds	NaOH aq. (pH 7.5)	DGS: extracting medium at ratio 1:82.81 (30 °C and 50 h)	Centrifugation and precipitation at pH 3.9	[42]
Watermelon seeds	Alkali (0.8%)	DGS: extracting medium at ratio 1:30 (40 °C and 30 min)	Precipitation at pH 4.5	[52]
Watermelon seeds	NaOH aq. (pH 12.0)	DGS: extracting medium at ratio 1:10 (w/v); stirring (1 h); 2 additional extractions after centrifugation	Precipitation at pH 4.0	[48,50]
Jujube seeds	(I) Water; (II) 50 mM Tris-HCl (pH 7.5); (III) 0.6 M NaCl (0.1% HCl); and (IV) acetic acid (5%)	(I and II) 1.5 g ground seeds and 20 mL solvent; stirring (4 °C and 2 h); (III) Idem at ground seeds:solvent 1:3 (w/v); and (IV) 5 g and 20 mL solvent. Shaking overnight (80 rpm)	(I, II, and III) Precipitation with $(NH_4)_2SO_4$ and dialysis (24 h, 4 °C); (IV) filtration and precipitation with acetone	[45]
Chinese cherry seeds	NaOH aq. (pH 10.0)	DGS: extracting medium at 1:20 (w/v); stirring (40 °C and 40 min)	Filtration and precipitation at pH 3.84	[55]
African breadfruit seeds	NaOH aq. (pH 9.0)	DGS: extracting medium at ratio 1:10 (w/v); stirring (30 min at room temperature)	Precipitation at pH 4.5	[56]
Pumpkin (*Cucurbita pepo*) seeds	NaOH aq. (pH 10.0)	DGS	Precipitation at pH 5.0	[57,61]
Pumpkin (*Cucurbita pepo*) oil cake	NaOH aq. (pH 10.0)	DGS	Filtration and precipitation at pH 5.0	[58,62]
Pumpkin (*Cucurbita pepo*) oil cake	NaOH aq. (pH 10.0)	Defatted oil cake: extracting media at ratio 1:10 (w/v)	Precipitation at pH 5.0	[59]
Pumpkin (*Cucurbita pepo*) oil cake	NaOH aq. (pH 11.0)	DGS: extracting medium at 1:30 (w/v); stirring (50 °C and 1.5 h)	Precipitation at pH 5.3	[60]
Pumpkin (*Cucurbita moschata*), watermelon, and bottle gourd seeds	Tris HCl (pH 8.0)	200 mg of DGS and 50 mL buffer (1 h)	Centrifugation and precipitation with acetone	[32]
Bottle gourd seeds	50 mM phosphate buffer (10 mM EDTA, 100 mM KCl, 1 mM DTT, and 1% SDS)	Sample: extracting medium at ratio 1:3 (w/v) (3 times)	Filtration and precipitation with chilled ethanol	[30]

Table 1. *Cont.*

Residue	Extracting Media	Extraction Conditions	Protein Purification Conditions	Refs.
Wax gourd seed	20 mM phosphate buffer (pH 6.5, 5.0 mM EDTA, and 10 mM DTT) (buffer I) and phosphate (2.0 mM EDTA, 1 mM DTT, urea 4 M, and 2% Triton X 100) (buffer II)	500 g of DGS and buffer I, 3 h; centrifugation and 2nd extraction under same conditions; centrifugation and 3rd extraction with buffer II and centrifugation	Dialysis, centrifugation, and filtration	[44]
Jujube seeds	Tris-HCl (pH 7.5)	1.5 g of ground seeds and 50 mM buffer (2 h)	Centrifugation, precipitation with $(NH_4)_2SO_4$ (4 °C), centrifugation, and dialysis	[46]
Milled rapeseed	50 mM Tris-HCl (pH 8.5, 750 mM NaCl, 5 mM EDTA, and 0.3% $Na_2O_5S_2$)	0.1 g/L (room temperature, 1 h)	Centrifugation	[23]
Pumpkin (*Cucurbita moschata*) seeds	Osborne method: (1) water; (2) Tris-HCl (5% NaCl); (3) isopropanol (55%); and (4) acetic acid (0.2 N)	150 mg of DGS and 10 mL extracting medium (60 min). Next extractions at extracting medium:sample ratio 7:1	(I, II, and III) Centrifugation and precipitation with acetone	[20]
Muskmelon seeds	Osborne method: (1) water; (2) NaCl (5%); (3) NaOH (0.1 M); and (4) ethanol (70%)	100 g of DGS and 500 mL extracting medium (60 min)	(I) Centrifugation; (II) centrifugation and dialysis; (III) precipitation at pH 4.0; and (IV) evaporation at 40 °C	[47]
Bottle gourd seeds	Osborne method: (1) water; (2) Tris-HCl (100 mM, pH 8.1, 0.5 M NaCl); (3) isopropanol (55%); and (4) acetic acid (0.2 N)	8 g of DGS and 60 mL extracting medium (60 min). Next extractions at extracting medium:sample ratio 7:1	(I, II, III, and IV) Centrifugation and precipitation with acetone	[31]
Watermelon seeds	Osborne method	-	-	[49]
ULTRASOUND-ASSISTED EXTRACTION				
Pomegranate peel	130 mM Tris-HCl (pH 7.5, 0.5%, SDS, and 0.25% DTT)	150 mg milled peel and 5 mL buffer using HIFU (30%, 1 min)	Evaporation and precipitation with cold acetone	[3]
Peach, plum, apricot, cherry, and olive seeds	100 mM Tris-HCl (pH 7.5, 0.5% SDS, and 0.5% DTT)	30 mg DGS and 5 mL buffer using HIFU (30%, 1 min)	Precipitation with cold acetone	[11,13]
Plum seeds	100 mM Tris-HCl (pH 7.5, 1% SDS, and 0.25% DTT)	30 mg DGS and 5 mL buffer using HIFU (30%, 1 min)	Precipitation with cold acetone and filtration	[14,35]
Olive and peach seeds	100 mM Tris-HCl (pH 7.5, 0.5% SDS, and 0.5% DTT)	30 mg DGS and 5 mL buffer using HIFU (30%, 5 min)	Precipitation with cold acetone	[16]
Olive seeds	125 mM Tris-HCl (pH 7.5, 1% SDS, and 0.1% DTT)	30 mg milled seeds and 5 mL buffer using HIFU (30%, 5 min)	Precipitation with cold acetone	[17]

Table 1. *Cont.*

Residue	Extracting Media	Extraction Conditions	Protein Purification Conditions	Refs.
Plum and peach seeds	50 mM Tris-HCl (pH 7.4) and 15 mM NaCl (buffer I)50 mM Tris-HCl (pH 7.4) and 15 mM NaCl and 1% SDS and 25 mM DTT (buffer II)	200 mg DGS and 10 mL of buffer I or II (10 min) and shaking (overnight)	Evaporation and precipitation with cold acetone	[29]
Cherry seeds	100 mM Tris-HCl (pH 7.5, 1% SDS, and 0.5% DTT)	30 mg DGS and 5 mL buffer using HIFU (30%, 5 min)	Precipitation with cold acetone	[33]
PRESSURIZED LIQUID EXTRACTION				
Pomegranate peel	Ethanol (70% (v/v))	2 g ground dried peels and 8 g sand (1500 psi; 120 °C; static extraction time, 3 min; extraction time, 12 min; static cycles, 1)	Evaporation and precipitation with cold ethanol	[25]
EXTRACTION USING DEEP EUTECTIC SOLVENTS				
Pomegranate peel	Choline chloride:AA:H_2O in 1:1:10 molar ratio	150 mg dried peels and 5 mL extracting medium (HIFU, 11 min, and 30%)	Evaporation and precipitation with cold ethanol	[25]
ULTRASOUND–MICROWAVE SYNERGISTIC EXTRACTION				
Pumpkin (*Cucurbita moschata*) seeds	PEG 200-choline chloride at 3:1 molar ratio	Microwave-assisted extraction (6 min, 120 W); ultrasound-assisted extraction (30 min, 240 W); water bath extraction (43 °C and 60 min); ultrasound–microwave synergistic extraction (28%, 28 g /L, 140 W, 43 °C and 4 min)	Isoelectric point precipitation; ethanol precipitation; centrifugation; centrifugation, isoelectric point precipitation, and ethanol precipitation	[64]
EXTRACTION USING PULSED ELECTRIC FIELD (PEF) AND HIGH VOLTAGE ELECTRICAL DISCHARGE (HVED)				
Olive seeds	Water	Sample: extracting medium at ratio 10 (w/w); pretreatment with HVED, PEF, and ultrasound; extraction (2 min and 150 rpm)	-	[34]
Rapeseed press-cake	Water	Sample: extracting medium at ratio 20 (v/w); HVED (240 kJ/kg)	-	[24]

Table 1. *Cont.*

Residue	Extracting Media	Extraction Conditions	Protein Purification Conditions	Refs.
Mango peels	I and II) Water; III) water (pH 11.0); IV) water and water (pH 11.0)	300 g of sample at ratio 1/10 (w/v); (I) PEF (13.3 kV/cm, 0.5 Hz); (II) HVED (40 kV/cm, 0.5 Hz); (III) aqueous extraction (20–60 °C and pH = 2.5, 6.0, 11.0); (IV) PEF; and (I) and aqueous extraction (50 °C, pH 6.0, and 3 h)	-	[26]
Papaya peels	(I and II) Water; (III) water (pH 11.0); (IV) water and water (pH 11.0)	300 g sample at ratio 1:10 (w/v); (I) PEF (13.3 kV/cm, 0,5 Hz); (II) HVED (40 kV/cm, 0.5 Hz); (III) aqueous extraction (20–60 °C, pH = 2.5, 6.0, 11.0); (IV) PEF; and (I) aqueous extraction (50 °C, pH 7.0, and 3 h)	-	[54]
METHODS USING NANOMATERIALS				
Plum seeds	100 mM Tris–HCl buffer (pH 7.5, 1% SDS, and 0.25% DTT)	30 mg DGS and 5 mL buffer using HIFU (30%, 1 min)	3G carboxylate-terminated dendrimers at pH 1.8 (30 min)	[74]
Plum seeds	100 mM Tris–HCl buffer (pH 7.5, 1% SDS, and 0.25% DTT)	30 mg DGS and 5 mL buffer using HIFU (30%, 1 min)	2G dimethylamino-terminated dendrimers at pH 7.5 (30 min)	[75]
Plum seeds	-	3G single wall carbon nanotubes functionalized with sulphonate-terminated carbosilane dendrimers at pH 7.5 with shaking (1 h)	Ultrafiltration	[76]
Peach seeds	-	2G gold nanoparticles coated with carbosilane dendrimers with carboxylate groups at pH 2.5 with shaking (2 h)	Ultrafiltration	[77]

Note: DGS: defatted ground seeds; Tris: tris(hydroxymethyl)aminomethane; DTT: dithiothreitol; SDS: sodium dodecyl sulfate; HIFU: high-intensity focused ultrasound.

4.3. Pressurized Liquid Extraction

Pressurized liquid extraction (PLE) uses temperatures and pressures in the ranges of 50–200 °C and 35–200 bar, respectively, to extract desired compounds. The high temperature enhances the solubility and mass transfer rate while reducing the viscosity and surface tension of solvents. The high pressure allows the solvent to rise above the normal boiling point temperature while keeping their liquid state. These conditions favor the penetration of solvents into the sample matrix and the analyte mass transfer. PLE requires a much lower amount of solvents than conventional solid–liquid extraction, and, in addition, it results in a higher yield and a reduced extraction time. Furthermore, PLE is usually carried out using environmentally friendly solvents such as water or ethanol [78]. Ethanol can be produced from raw agricultural materials (cereals and sugar beet) and from waste and residues (straws). Though PLE has been mostly applied to the extraction of small molecules, it can also be useful in protein extraction, as shown in previous works devoted to the extraction of algae proteins [79,80].

In the case of fruit residues, this technique has only been used for the extraction of proteins from pomegranate peels [25]. In order to achieve the highest extraction yield, different parameters were optimized: extracting the solvent, temperature, static time, and the presence of additives in the solvent. The highest extraction yield was obtained when using 70% (v/v) ethanol as the extracting solvent and a high temperature (120 °C). No significant differences were observed by increasing the extraction time over 3 min, repeating cycles, or adding DTT or urea to the extracting solvent. Under optimized conditions, the extraction yield (9 mg of proteins/g of pomegranate peel) was lower than that obtained using HIFU [3] (15 mg/g). Further studies enabled researchers to observe that the proteins extracted with every technique were different and that both techniques could be complementary to obtain a more comprehensive extraction of proteins from pomegranate peels [25].

Moreover, the use of non-toxic, cheap, and environmentally friendly solvents in small quantities, as well as its easy automation, make this technique a great candidate for food industry applications [73]. Nevertheless, working with ethanol at the industrial scale requires certain safety conditions.

4.4. Extraction Using Deep Eutectic Solvents

Deep eutectic solvents (DES) are sustainable extractants that are usually derived from renewable resources [81]. DES are obtained by mixing two solid organic compounds, a hydrogen-bond acceptor (HBA), such as quaternary ammonium salts, and a hydrogen-bond donor (HBD), such as amides, alcohols, and acids, at an appropriate molar ratio (eutectic composition). HBD and HBA associate with each other by means of hydrogen bond interactions [25]. DES formation is based on a phenomenon called freezing point depression. The result of that mixing is a liquid solvent at relatively low temperatures. Choline chloride is the most used HBA, while urea, citric acid, glucose, tartaric acid, succinic acid, and glycerol are other usual HBDs employed in the synthesis of DES [81–83]. DES have been mostly used in the extraction of small compounds, although some DES have been applied in the extraction of some standard proteins (bovine serum albumin, papain, and wheat gluten) [54,55] and in the extraction of proteins from brewers spent grain [84].

More recently, DES have been employed for the extraction of proteins from pomegranate peels [25]. Different DES were synthesized, and their extraction capacities were compared. Choline chloride and sodium acetate were employed as HBAs, while different HBDs were tried (ethylene glycol, glycerol, acetic acid, glucose, sorbitol, and acetic acid). A choline chloride:acetic acid DES was selected as the most effective. The acceleration of extraction was possible using HIFU at 60% for 11 min. Under these conditions, DES extracted 20 mg of proteins/g of pomegranate peel. A comparison with results obtained for the same sample with HIFU and PLE showed that DES had a higher protein extraction capability (HIFU (15 mg of proteins/g of pomegranate peel) [3] and PLE (9 mg proteins/g of pomegranate peel) [25].

While DES seem to comprise a promising group of green solvents, further research is needed in order to confirm whether they are nontoxic to animals and the environment [85]. Moreover, its applicability at the industrial scale is, so far, compromised by the high price of DES. Further studies on the possible reutilization of DES are very interesting in this regard.

4.5. Microwave-Assisted Extraction

Microwave-assisted extraction (MAE) uses microwave radiation to favor extraction processes. The overall effect is the warming of the sample due of the dissipation of the radiation energy by thermal conduction, in the case of ions, or by the rotation of dielectric dipoles and friction with the solvent, in the case of polar molecules. MAE has been widely employed for the extraction of small molecules and, to a lesser extent, for the extraction of proteins [86,87].

The extraction of proteins from pumpkin (*Cucurbita moschata*) seeds was performed using MAE and different DES as extracting solvents [64]. Poly (ethylene glycol) (PEG 200) was employed as the HBD, while several HBAs were tried. A PEG 200:choline chloride mixture at a 3:1 ratio was selected. The performance of MAE was compared with results obtained by using a conventional water bath extraction, the extraction using a HIFU probe, and the extraction using both MAE and HIFU, using the same DES as the extracting medium. Ultrasound–microwave synergistic extraction showed a better average extraction efficiency (94%) with less solvent and shorter extraction times (4 min) than water bath extraction (1 h), MAE (6 min), and HIFU (30 min).

Though this technique can be scaled-up, its application is complex due to the non-homogenous depth of microwave radiation that results in a non-uniform sample heating. Moreover, it is necessary to work under proper conditions to avoid potential hazards.

4.6. Extraction Using Pulsed Electric Field (PEF) and High Voltage Electrical Discharge (HVED)

The application of electric discharges has been used in the food industry to increase food shelf life since it has the ability to inactivate microorganisms. The technique is based on the application of electric discharges that create currents and bubbles (cavitation bubbles) that expand and implode, thus causing pressure variations of up to 100 bar. These pressure changes allow for the permeation of cell walls in a controlled way, unlike classical treatments in which tissue structure is entirely disrupted, thus losing its selectivity and becoming permeate to all intracellular compounds. This technique has a high potential in the extraction of proteins, although it has been scarcely used for this purpose. Moreover, it allows for work at low temperature conditions, which can also be an advantage over other techniques that require the use of high temperatures. Indeed, non-thermal emerging techniques have been proposed to shorten the processing time, to increase recovery yield, to control Maillard reactions, to improve product quality, and to enhance extract functionality [34,88].

Extraction using pulsed electrical field (PEF) is a non-thermal treatment of very short duration (from several nanoseconds to several milliseconds) that consists of the application of pulses with amplitudes from 100–300 V/cm to 20–80 kV/cm. This method induces the permeabilization of biological membranes by electrical piercing, which is called electroporation. The cell network maintains its capacity to act as a barrier for the passage of some undesired compounds, which improves extraction selectivity. Furthermore, plant materials treated with PEF seem to be less altered than thermally treated ones [54]. However, there are two main problems for its industrial application: the non-uniform nature of the ideal distribution of electric pulses and the limited variety of suitable solvents [73].

High voltage electrical discharge (HVED)-technology has also been recently studied for enhancing the extraction of bioactive compounds from different raw materials. HVED leads to the generation of hot, localized plasmas that emit high-intensity UV light, produce shock waves and bubble cavitation, and generate hydroxyl radicals from water photo-dissociation. In general, this technique provides a higher extraction rate than both PEF and ultrasound, but it may produce contaminants such as chemical products from electrolysis and free reactive radicals [88]. Some research has been published trying to scale up this technique to the pilot scale, but further research is still needed in order to achieve its industrial application [89].

These methods have been applied, at the lab scale, in the extraction of proteins from rapeseed press-cake [24], olive seeds [34], papaya seeds [54], and mango peels [26]. In some examples, they were employed as extraction techniques, while, in other cases, they were used as pretreatments followed by a solid–liquid extraction.

Proteins from mango peels were extracted by Parniakov et al. [26] using HVED and PEF, and the results were compared with those obtained by conventional aqueous extraction. HVED showed a higher protein yield than PEF and conventional aqueous extraction. Results obtained by conventional extraction improved when mango peels were pretreated with HVED and, especially, after PEF pretreatment. Very similar results were obtained when applying PEF and HVED for the extraction of proteins from papaya seeds [54]. Again, the protein yield obtained by conventional solid–liquid extraction significantly improved when the sample was pretreated with PEF.

A different approach was followed for the extraction of proteins from olive seeds [34]. In this case, HVED and PEF were used as pretreatments, followed by extraction with an aqueous:ethanol solution at different pHs. Extraction after HVED, using 23% ethanol at pH 12.0, provided the highest yield. Higher ethanol concentrations generated a smaller yield, probably due to the aggregation of proteins. In the case of the rapeseed press-cake [24], HVED was employed as the only step, and no subsequent solid–liquid extraction was performed. In this case, the use of powers higher than 240 kJ/kg did not improve the protein extraction yield. This fact was attributed to the release of oxygen reactive species that could react with proteins.

4.7. Extraction and Purification Using Nanomaterials

The use of nanomaterials in the extraction and purification of proteins is an interesting strategy to increase the sustainability of these steps. Nanomaterials present, at least, one of their dimensions in the nanoscale (1–100 nm). This is associated with extraordinary mechanical properties and enhanced electrical, magnetic, optical, thermal, or chemical properties. Moreover, they usually have good reactivity and can be easily functionalized. Different nanomaterials have been employed in the extraction and purification of proteins [90], although they have been scarcely applied in the case of fruit residues.

Dendrimers are a kind of nanomaterials with a structure similar to tree roots. They consist of layers called generations in which functional groups can be introduced. Carbosilane dendrimers are a special type of dendrimer that contain silicon atoms, have high stability and biocompatibility, and are easily functionalized. Carbosilane dendrimers can interact with proteins, and they have been applied for the purification and extraction of proteins from peach and plum seeds [74,75].

The use of carbosilane dendrimers functionalized with carboxylates groups under acid conditions has been found to result in the precipitation of proteins. This ability has enabled the development of a method for the purification of plum proteins as an alternative to acetone precipitation. The best yield was obtained with third generation dendrimers [74]. On the other hand, the ability of different cationic carbosilane dendrimers functionalized with amino, trimethylammonium, or dimethylamine groups to interact with proteins has also been studied. Second generation carbosilane dimethylamine-terminated dendrimers were selected for the purification of plum proteins. They resulted in the precipitation of 97% of proteins that could be precipitated with cold acetone [75]. The same authors, in other work, used single-wall carbon nanotubes coated with carbosilane dendrimers that were functionalized with sulphonate groups to extract proteins from plum seeds. Protein extraction yield was similar to the obtained with HIFU [76].

Moreover, gold nanoparticles functionalized with carbosilane dendrimers have been employed in the extraction of proteins from peach seeds [77]. Gold nanoparticles coated with dendrimers functionalized with sulphonate, carboxylate, or trimethylammonium groups were used and compared. The highest recovery of proteins was obtained with gold nanoparticles coated with second-generation carbosilane dendrimers functionalized with carboxylate groups at acid pH. Nevertheless, the recovery of proteins was low, and the strength of protein–dendrimer interactions was so high that very harsh conditions were required for their disruption.

5. Methods Used for the Release of Antioxidant Peptides

Once proteins have been extracted from a fruit residue, they must be submitted to a step to release peptides. There are three main approaches for this purpose: microbial fermentation with proteolytic microbes, proteolysis using enzymes from plants and microorganisms, and proteolysis using gastrointestinal enzymes. Table 2 shows the conditions employed to obtain peptides with antioxidant properties from fruit residues.

Table 2. Conditions employed to obtain antioxidant peptides from fruit residues.

Fruit Residue	Enzyme/Microorganisms	Buffer (pH)	Temperature (°C)	Time (h)	Refs.
	RELEASE OF PEPTIDES BY MICROBIAL FERMENTATION				
Tomato seed	*Lactobacillus plantarum*	—	37	24	[36]
Tomato seed	Water kefir microbial mixture	—	37	24	[37,41]
Tomato seed	*Bacillus subtilis*	—	40	20	[38]
Tomato seed	*Bacillus subtilis*	—	37	24	[39,40]
	RELEASE OF PEPTIDES BY ENZYMATIC DIGESTION				
Pumpkin oil cake	Alcalase	Tris-HCl (0.1 M and pH 8.0)	50	1	[58]
	Flavourzyme			1	
	Alcalase and flavourzyme			2	
Pumpkin oil cake	Alcalase	Phosphate (pH 8.0)	50	0–2.5	[62]
	Flavourzyme	Phosphate (pH 7.0)	50		
	Pepsin	Phosphate (pH 3.0)	37		
Pumpkin oil cake	Alcalase	Tris-HCl (pH 9.0)	50	3.5	[59]
	Trypsin	Tris-HCl (pH 8.0)	35	5	
Pumpkin seed	Acid protease	pH 2.5	50	5	[60]
	Alcalase	pH 8.0	55		
Pumpkin meal	Flavourzyme	pH 7.0	50	5	[63]
	Protamex	pH 6.5	50		
	Neutrase	pH 7.0	50		
Peach, plum, apricot, and olive seeds	Pepsin and pancreatin	pH 2.0 and pH 8.0	37	3	[11]
Apricot seeds	Alcalase	Borate (5 mM and pH 8.5)	50	4	[11]
	Thermolysin	Phosphate (5 mM and pH 8.0)		4	
	Flavourzyme	Phosphate (5 mM and pH 7.5)		8	
Plum seeds	Alcalase	Borate (5 mM and pH 8.5)	50	3	[14,35]
	Thermolysis	Phosphate (5 mM and pH 8.0)	50	4	
	Flavourzyme	Phosphate (5 mM and pH 7.0)	50	7	
	Protease P	Phosphate (5 mM and pH 7.5)	40	24	
Apricot seeds	Alkaline and flavor proteases	-	-	-	[27]
Peach seeds	Alcalase	Phosphate (5 mM and pH 8.0)	50	4	[28]
	Thermolysin	Phosphate (5 mM and pH 8.0)	50	4	
	Flavourzyme	Ammonium bicarbonate (5 mM and pH 6.5)	50	3	
	Protease P	Phosphate (5 mM and pH 7.5)	40	7	

Table 2. *Cont.*

Fruit Residue	Enzyme/Microorganisms	Buffer (pH)	Temperature (°C)	Time (h)	Refs.
Cherry seeds	Alcalase	Borate (pH 8.5)	50	7	[33]
	Thermolysin	Phosphate (pH 8.0)			
	Flavourzyme	Bicarbonate (pH 6.0)			
Chinese cherry seeds	Alcalase and Neutrase	Water (pH 7.5)	50	2	[55]
Olive and peaches seeds	Alcalase	Borate (5 mM and pH 8.5)	50	4	[16]
Olive seeds	Alcalase	Phosphate (5 mM and pH 8.0)	50	2	[17]
	Thermolysin	Phosphate (5 mM and pH 8.0)			
	Flavourzyme	Ammonium bicarbonate (5 mM and pH 6.0)			
	Trypsin	Tris-HCl (5 mM and pH 9.0)			
	Neutrase	Phosphate (5 mM and pH 7.0)			
Tomato seeds	Alcalase	Phosphate (pH 8.0)	50	0.5–3	[43]
Tomato seeds	Alcalase	Phosphate (pH 8.0)	50	2.3	[21]
Milled rapeseed	Pepsin	Phosphate (0.1 M and pH 2.0)	40	3	[23]
	Trypsin	Phosphate (0.1 M and pH 7.0)	40	3	
	Alcalase	Phosphate (0.1 M and pH 7.0)	50	3	
	Subtilisin	Phosphate (0.1 M and pH 8.0)	60	3	
	Thermolysin	Phosphate (0.1 M and pH 8.0)	60	24	
Jujube seeds	Papain	Tris-HCl (50 mM and pH 6.5–7.5)	65	1.5	[45,46]
	Alcalase	Tris-HCl (50 mM and pH 6.5–8.5)	60		
	Protease P	Tris-HCl (50 mM and pH 7.5)	37		
Muskmelon seeds	Pepsin and Trypsin	pH 2.0 and pH 7.0	37	6	[47]
Pumpkin (*Cucurbita moschata*), watermelon, and bottle gourd seeds	Trypsin	Tris-HCl (50 mM and pH 7.5)	-	4	[32]
Watermelon seeds	Alcalase	Phosphate (5 mM and pH 8.0)	60	5	[48,50]
	Trypsin	Phosphate (5 mM and pH 8.0)	37		
	Pepsin	Glycine (5 mM and pH 2.2)	37		
Watermelon seeds	Alcalase	pH 8.5	55	3	[51]
	Papain	—	—	—	
	Pepsin	pH 2.4	37	3	
Watermelon seeds	Protease	—	—	—	[52]
	Pancreatin	—	—	—	
	Trypsin	—	—	—	
	Chymotrypsin	—	—	—	
Watermelon seeds	Alcalase	NaOH aq. (pH 9.0)	50	0.8	[53]
African breadfruit seeds	Trypsin, pepsin, and pancreatin	Water	-		[56]
Date palm seeds	Alcalase	pH 8.0	50	1	[65,66]
	Flavourzyme	pH 7.0		2	
	Thermolysin	pH 8.0		3	

Table 2. *Cont.*

Fruit Residue	Enzyme/Microorganisms	Buffer (pH)	Temperature (°C)	Time (h)	Refs.
Pomegranate peel	Alcalase	Borate (5 mM and pH 9.0)	50	2	[3]
	Thermolysin	Phosphate (5 mM and pH 7.5)	70	1	
Pomegranate peel	Alcalase Thermolysin	Borate (5–10 mM and pH 9.0) Phosphate (5 mM and pH 7.5) or borate (100 mM and pH 7.5)	50 70	2 1	[25]

Note: Enzymes in bold characters are the ones which yielded the higher quantity of antioxidant peptides in each research work.

Commercial proteases can be expensive, and their industrial application might not be economically efficient. Microbial fermentation is a more environment-friendly and cost-effective proposal that has been tested in tomato seeds [36–41]. Different microorganisms were chosen: *Bacillus subtilis*, *Lactobacillus plantarum*, and a mixture of microorganisms from kefir culture. *B. subtilis* has been traditionally used to obtain fermented soybean products, while *L. plantarum* is a lactic acid bacteria that has been widely used in food production and preservation. In these cases, no previous extraction of proteins was carried out, and ground seeds were directly added to the culture. In another case, a kefir culture containing different lactic acid bacteria and yeasts was employed [37,41]. In all cases, fermentations took place at 37–40 °C and required very high times ranging from 20 to 24 h.

Alternatively, commercial proteases such as Alcalase, thermolysin, flavourzyme, protease P, Neutrase, trypsin, papain, and pepsin have been preferred in most works. Alcalase (the commercial name of subtilisin Carlsberg endopeptidase) is the most used enzyme for obtaining antioxidant peptides. This is not surprising, because Alcalase is a very cost effective food-grade protease with a low specificity that enables the release of a wide range of short peptides [43]. Trypsin, conversely, is a highly selective protease and has not been very effective in antioxidant peptide releasing [90]. Other food-grade enzymes such as pepsin, papain, pancreatin, and flavourzyme have also resulted in hydrolysates with high antioxidant activity [28,32,33,45,46,52,65,66]. Hydrolysis times were much lower than the required in microbial fermentation. Indeed, they usually ranged from, one-to-five hours, although there are some procedures that took longer. Moreover, hydrolysis must be carried out at a controlled pH and temperature conditions to obtain an optimum hydrolysis degree, which usually involves the careful optimization of these parameters.

Different strategies have been employed for the release of bioactive peptides from pumpkin oil cake proteins. Vaštag et al [58] compared the capacity of two different enzymes (Alcalase and flavourzyme) for this purpose. Alcalase showed a higher capacity to release peptides from pumpkin oil cake than flavourzyme (a hydrolysis degree (DH) of 53% for Alcalase and 37% for flavourzyme) and resulted in a hydrolysate with a higher antioxidant capacity. The DH increased up to 69% when the hydrolysate obtained using Alcalase was further hydrolyzed with flavourzyme, although antioxidant activity decreased [58]. Other enzymes employed for the release of peptides from pumpkin oil residues are pepsin, trypsin, protamex, and Neutrase [59,60,62,63]. However, many authors have pointed out that Alcalase is the most promising protease to produce pumpkin protein hydrolysates with an improved nutritional quality, but flavourzyme was the best to obtain antioxidant peptides. More recently, Nourmohammadi et al. [59] compared the feasibility of Alcalase to release antioxidant peptides with that of the trypsin enzyme. Again, Alcalase was preferred. Alternatively, other authors proposed the use of a protease working at an acid pH to obtain highly antioxidant peptides from pumpkin seeds [60].

Seeds from *Prunus* fruits (plum, peach, cherry, and apricot) also present a high protein content and have been exploited to obtain antioxidant peptides [11,14,27–29,33,35,55]. Like in other samples, different enzymes (Alcalase, flavourzyme, protease P, and thermolysin) were tried, and the highest antioxidant activity was observed in the hydrolysates obtained with Alcalase and thermolysin. Hydrolysis conditions were optimized for every enzyme, and although hydrolysis times from 2 to 4 h were the most common, some enzymes such as protease P required times up to 24 h. Moreover, the capacity of these seeds to release antioxidant peptides after simulated gastrointestinal digestion with pepsin and pancreatin was also studied [11], and the results demonstrated that peptides were less antioxidant than those obtained with previous enzymes. Moreover, the residue obtained after cherry wine fermentation, which was mainly constituted by cherry seeds, was simultaneously hydrolyzed with the Alcalase and Neutrase enzymes [55]. The resulting hydrolysate showed high antioxidant properties.

Seeds from other fruits such as olive, muskmelon, watermelon, tomato, bottle gourd, African breadfruit, and milled rapeseed were also used as sources of antioxidant peptides [16,17,21,23,32,43,47,48,50]. Different enzymes have been employed, but Alcalase and pepsin have generally reported the highest hydrolysis degree and antioxidant activity. In fact, the hydrolysis degree and antioxidant activity of hydrolysates obtained from olive seed proteins using the Alcalase enzyme was higher than the obtained with thermolysin, flavourzyme, Neutrase, and Trypsin [17]. Alcalase digestion also showed the highest hydrolysis degree and antioxidant activity in milled rapeseed proteins, as compared to pepsin, trypsin, subtilisin, and thermolysin [23]. Unlike them, watermelon seeds showed the highest hydrolysis degree when using the pepsin enzyme, while trypsin, Alcalase, papain, protease, pancreatin, and chymotrypsin yielded a lower hydrolytic activity [50,52]. Osukoya et al. compared the capacity to release peptides from the African breadfruit seed of pepsin enzyme with that of trypsin and pancreatin. In this case, pancreatin was the enzyme that resulted in the highest amount of peptides [56].

Jujube (also called red or Chinese date) and date palm also contain seeds with high protein contents that have been evaluated as sources of antioxidant peptides. Ambigaipalan et al. [65,66] employed different combinations of Alcalase, thermolysin, and flavourzyme enzymes for the hydrolysis of date palm seed proteins. They observed that the combination of Alcalase and flavourzyme resulted in the hydrolysate with the highest antioxidant activity [65]. In other research works, papain was employed for the hydrolysis of jujube seed proteins, and the results were compared with the obtained when using Alcalase and protease P; the hydrolysate obtained using papain yielded the highest antioxidant activity [45,46].

Hernández-Corroto et al. optimized the hydrolysis of pomegranate peel proteins using Alcalase and thermolysin, and they evaluated the potential of these proteins to release antioxidant peptides. No significant difference was observed in the antioxidant activity of the hydrolysates obtained by both enzymes [3,25]. Moreover, further studies revealed the contribution of phenolic compounds coextracted with proteins to the antioxidant activity that was observed in a hydrolysate obtained with thermolysin.

The performance of enzymatic hydrolysis can be improved by the previous treatment of proteins with ultrasound. Indeed, Wen et al. demonstrated that ultrasound treatment had a significant impact on proteins structure by improving their susceptibility to hydrolysis [51,53]. They submitted watermelon seed proteins to three different ultrasound treatments—single (20 kHz), dual (20 kHz/28 kHz), and tri-frequency (20/28/40 kHz))—before hydrolysis with Alcalase. All ultrasound treatments increased protein hydrophobicity by causing changes in secondary structure of proteins. Under most effective treatment (dual frequency ultrasound), the degree of hydrolysis and antioxidant activity increased [53].

6. Evaluation of Antioxidant Activity of Peptides

Antioxidant activity, in a biological system, may occur through different mechanisms of action: (i) the inhibition of generation or the scavenging of ROS and reactive nitrogen species (RNS); (ii) the reduction of oxidants; (iii) the chelation of metals; (iv) as an antioxidant enzyme; and (v) the inhibition of oxidative enzymes [91]. Therefore, a comprehensive estimation of in vitro antioxidant activity must involve the use of different assays based on different mechanisms. Table 3 groups the assays employed for the evaluation of antioxidant activity in hydrolysates obtained from fruit residues.

Table 3. Assays employed for the evaluation of antioxidant activity.

Assay	Methodology	Refs.
EVALUATION OF THE CAPACITY TO SCAVENGE FREE RADICALS AND OXIDANT SPECIES		
Scavenging effect on hydrogen peroxide (H_2O_2) radicals	Measurement of the reduction in the absorbance of a H_2O_2 solution at 230 nm after incubation with potential antioxidants.	[20,31,48]
Scavenging effect on ABTS (2,2'-azino-bis(3-ethylbenzothiazoline-6-sulfonic acid) radicals	ABTS radicals that absorb at 734 nm are produced by the reaction of ABTS with potassium persulfate. The method evaluates the reduction in the absorbance of ABTS radicals due to the presence of potential antioxidants.	[3,11,14,16,17,20,24,25,28,31,33–37,41,42,45,46,49,51,52,54,55,58,62,65]
Scavenging effect on nitric oxide (NO) radicals	Nitric oxide radicals are formed from nitroprusside and the incubation of formed nitric oxide radicals with a Griess reagent (1% sulphanilamide, 2% H_3PO_4, and 0.1% naphthylethylene diamine dihydrochloride) results in nitrite ions. Nitrite ions can be measured by the formation of a compound that absorbs at 546 nm. The scavenging of nitric oxide radicals by potential antioxidants reduces nitrite ion formation and their absorbance at 546 nm.	[20,31,49]
Scavenging effect on DPPH (1,1-diphenyl-2-picrylhydrazyl) radicals	Measurement of the decrease in the absorption of DPPH radicals at 515–517 nm when potential antioxidants are added.	[14,17,20,21,26,27,30,31,34,36–49,51–56,59–61,63,65]
EVALUATION OF THE CAPACITY TO INHIBIT OXIDATION REACTIONS		
Inhibition of formation of superoxide (O^{2-}) radicals	The assay measures the rate of pyrogallol autooxidation in presence and absence of potential antioxidants at 320–420 nm.	[50,55]
Inhibition of formation of hydroxyl (OH•) radicals	Hydroxyl radicals are generated by the oxidation of Fe^{2+} to Fe^{3+} in the presence of H_2O_2. The presence of Fe^{2+} is monitored by the formation of a complex with 1,10-phenanthroline that absorbs at 536 nm. The presence of potential antioxidants inhibits the oxidation of Fe^{2+} and results in an absorbance increase.	[3,11,14,16,17,25,27,28,33,35,47,55,65]
Oxygen radical antioxidant capacity (ORAC)	The method is based on the oxidation of fluorescein by reactive oxygen species (ROS) resulting from the radical initiator 2,2'-azobis(2-methylpropionamidine) dihydrochloride. The inhibition of fluorescein oxidation by the presence of potential antioxidants is measured from the increase in fluorescence intensity.	[51,65]
EVALUATION OF THE REDUCING POWER		
Ferric reducing antioxidant power (FRAP)	Measures the ability of potential antioxidants to reduce Fe^{3+} from the ferricyanide complex to Fe^{2+}-complex. Formation of Fe^{2+}-complex is measured at 700 nm.	[14,16,20,27,28,30,31,33,35,38–40,45–47,49,50,52,56,58,62,63,65]
Ammonium phosphomolybdenum	The method evaluates the capacity of potential antioxidants to reduce Mo^{6+} to Mo^{5+}. Presence of Mo^{5+} is monitored by the subsequent formation of a green phosphor/Mo^{5+} complex that absorbs at 695–65 nm.	[20,21,30,31,59]

Table 3. *Cont.*

Assay	Methodology	Refs.
EVALUATION OF THE METAL QUELATION ACTIVITY		
Ferrous ion chelation activity (FICA)	Ferrozine reacts with Fe^{2+} to form a complex that absorbs at 562 nm. In the presence of chelating agents, the complex is disrupted, resulting in a decrease in absorption at 562 nm.	[45,46,52,53,59,61,63,65]
Cuprous ion chelation activity (CICA)	Reaction of pyrocatechol and Cu^{2+} results in a substance that absorbs at 632 nm. The presence of a metal chelator disrupts this molecule and reduces the absorbance.	[47]
EVALUATION OF THE CAPACITY TO INHIBIT LIPIDS AND LIPOPROTEINS OXIDATION		
Ferric thiocyanate	Primary products resulting from the oxidation of linoleic acid are incubated with EtOH, ammonium thiocyanate, and $FeCl_2$, leading to the formation of $Fe(SCN)^{2+}$ that absorbs at 500 nm. Presence of potential antioxidants results in the inhibition of linoleic acid oxidation and the reduction of absorption.	[14,16,17,28,33,35,44,49,56,63]
Thiobarbituric acid reactive substances (TBARS)	The presence of secondary oxidation products formed during oxidation of linoleic acid is measured by the reaction of one of them, the malondialdehyde, with SDS, acetic acid, and TBA at 532 nm. The presence of potential antioxidants reduces this absorbance.	[20,23,31,32,44,53,57,66]
β-carotene linoleate	It measures the ability of potential antioxidants to decrease the oxidative bleaching of β-carotene in an oil-in-water emulsion. The reaction is monitored by measuring the absorbance at 470 nm immediately after the addition of a potential antioxidant.	[66]
Inhibition of Cu^{2+}-induced low-density lipoprotein (LDL) peroxidation	This assay measures the peroxidation induced by cupric sulfate in LDL. Presence of potential antioxidants results in the inhibition of the oxidation and the reduction of the absorbance of conjugated dienes at 344 nm.	[66]
EVALUATION OF THE CAPACITY TO INHIBIT DNA OXIDATION		
Supercoiled-to-Nicked-Circular-Conversion (SNCC)	Oxidation of supercoiled DNA into nicked circular DNA in the presence of Cu^{2+} and H_2O_2 is monitored by measuring the fluorescent intensity of ethidium-stained nicked circular DNA. The presence of a potential antioxidant inhibits this reaction, and the signal corresponding to the oxidized form of DNA decreases.	[30]
Inhibition of peroxyl and hydroxyl radical-induced supercoiled strands scission	Strand scission of supercoiled DNA is measured in the presence of peroxyl and hydroxyl radicals. After incubation, DNA is separated by gel electrophoresis, and the intensity of supercoiled DNA bands in the presence and absence of potential antioxidant are compared.	[66]

Table 3. *Cont.*

Assay	Methodology	Refs.
	EVALUATION OF THE CAPACITY TO INHIBIT OXIDATIVE DAMAGE INDUCED IN CELLS	
2′, 7′-dichloro-dihydro-fluorescein diacetate (DCFH-DA) fluorescent probe	Oxidative stress in cells is induced by the addition of a strong oxidant (H_2O_2 or other peroxide). DCFH-DA fluorescence probe, added to cell culture, reacts with ROS to produce fluorescent DCF that is measured at an λexcitation of 488 nm and an λemission of 585 and 530 nm. The presence of a potential antioxidant inhibits ROS generation and DCF signal decreases.	[16,51]
Intracellular concentration of Ca^{2+} determination	Intracellular Ca^{2+} is measured with fluorescent dye Fura-2 AM. Fura-2AM is cleaved by intracellular esterase, and the resulting Fura-2 can bind to Ca^{2+} and cause strong fluorescence under a 330–350 nm excitation light. Fluorescence intensity decreases in H_2O_2-damaged cells treated with potential antioxidants.	[51]
Acridine orange/ethidium bromide (AO/EB) fluorescent staining	Cell membrane damage is measured by evaluating the staining of DNA with EB or AO using an inverted fluorescence microscope. The presence of potential antioxidants will reduce the number of red cells resulting from the staining with EB and will increase the number of green cells resulting from the staining with AO.	[51]

Assays based on the scavenging of free radicals or oxidants have been the most commonly used to evaluate the antioxidant activity of hydrolysates. The two most common radical scavenging assays used non-biological oxidant species: ABTS (2,2′-azino-bis(3-ethylbenzothiazoline-6-sulfonic acid) and DPPH (1,1-diphenyl-2-picrylhydrazyl) [92] radicals. Assays using biological species such as nitric oxide and hydrogen peroxide have been less employed. All these assays consist of the spectrometric monitoring of their own radical or of a derived compound. Antioxidant capacity is estimated from the reduction in the absorbance of the oxidant or derived compound in the presence of potential antioxidants. A DPPH assay was employed to evaluate the antioxidant activity of peptides obtained by fermentation from tomato seeds. Fermentation with *B. subtilis* for 20 h resulted in an increase in DPPH activity from 23.5 to 68.5% [38]. Similarly, fermentation with kefir culture or *L. Plantarum* enabled a decrease from 41.24 and 40.89 to 10.84 and 4.95 μL/mg, respectively, in the IC_{50} values obtained by the DPPH radical scavenging assay [36,37].

Other kind of methods have evaluated the capacity of potential antioxidants to inhibit an oxidation reaction. In some cases, these reactions result in the formation of free radicals such as superoxide radicals (O^{2-}) or hydroxyl radicals (OH•) [3,11,14,16,17,25,27,28,33,35,47,55,65,66]; in others, by using different radicals (ROO•, HO•, O_2•), a fluorescent probe is oxidized [51,65]. Superoxide radicals are formed by the oxidation of pyrogallol at basic pH. The antioxidant activity of the hydrolysates obtained from watermelon and Chinese cherry seeds was estimated by evaluating their capacity to inhibit this reaction and, thus, the formation of superoxide radicals [50,55]. On the other hand, hydroxyl radicals are generated by the Fenton reaction that is based on the oxidation of Fe^{2+} to Fe^{3+} in presence of H_2O_2. The formation of hydroxyl radicals is inhibited by the presence of antioxidant peptides from *Prunus* fruit, date, muskmelon, and olive seeds and from pomegranate peels [3,11,14,17,25,27,28,33,35,47,55,65,66]. A last assay in this category is based on the oxidation of fluorescein by the presence of ROS. The inhibition of this oxidation reaction is possible with peptides obtained from watermelon and date seeds [51,65].

Two methods have been employed to evaluate the reducing power of potential antioxidants in fruit residues: the ferric reducing antioxidant power assay (FRAP) and the ammonium phosphomolybdenum assay. Both are based on monitoring the capacity of peptides to reduce a probe cation (Fe^{3+} or Mo^{6+}) [14,16,20,21,27,28,30,31,33,35,38–40,45–47,49,50,52,56,58,59,63,65]. The FRAP method, based on the reduction of Fe^{3+}, has been the most employed by far.

Metal chelation activity has been explored in ferrous (FICA) and cupric (CICA) ion chelation activity assays. Both assays are based on the spectrometric measurement of colored complexes of Fe^{2+} and Cu^{2+} with ferrozine and pyrocatechol, respectively. The chelation of these metal ions by potential antioxidants avoids the formation of these colored complexes. These assays have been employed to evaluate the antioxidant activity of peptides released from watermelon, jujube, pumpkin, and date seeds [45–47,52,53,59,61,63,65].

The oxidation of biological molecules (lipids, proteins, and DNA) can result in the development of chronic diseases. For example, the oxidation of low density lipoproteins (LDL) cholesterol can result in atherosclerotic lesions, and the oxidation of DNA plays an essential role in the development of cancer. Moreover, the presence of ROS can result in the oxidative degradation of lipids through a process called lipid peroxidation, which can cause serious cell damage. Therefore, some antioxidant assays have been based on evaluating the capacity of peptides to inhibit oxidative damages on these molecules.

The lipid peroxidation inhibition of peptides and proteins from fruit residues has been measured using the ferric thiocyanate (FTC) assay and the thiobarbituric acid reactive substances (TBARS) assay. The FTC assay measures the primary products (hydroperoxides) formed during the oxidation of a fatty acid (e.g., linoleic acid or oleic acid). Hydroperoxides oxidize Fe^{2+} to Fe^{3+}, and the latter forms a colored ferric thiocyanate complex. This method has been employed to evaluate the antioxidant activity in peptides obtained from *Prunus* fruits, African breadfruit, pumpkin, and olive seeds, as well as in proteins from watermelon seeds. The TBARS assay, on the other hand, measures a secondary product formed during lipid peroxidation (malondialdehyde) by its reaction with thiobarbituric acid. This assay was employed to evaluate the antioxidant activity in peptides released from pumpkin

(*Cucubirta pepo* and *Cucurbita moschata*), watermelon, and wax gourd seeds [20,44,53,57]. The TBARS assay was also employed to evaluate the protection against oxidants of date seed protein hydrolysates using a biological model system—a cooked comminuted salmon. Hydrolysates obtained using Alcalase and flavourzyme and flavourzyme and thermolysin were able to achieve TBARS inhibition values of 32% and 30%, respectively, after seven days of storage compared with a positive control (butylated hydroxytoluene), which only achieved a 7% inhibition.

Two different assays, based on the inhibition of the oxidation of lipoproteins, have been employed to evaluate the antioxidant activity of peptides released from date seeds. One of them was based on the inhibition of β-carotene oxidation, and the other was based on the inhibition of low-density lipoprotein peroxidation. Date seed proteins were hydrolyzed using different combinations of Alcalase, thermolysin, and flavourzyme. The hydrolysate obtained with flavourzyme and thermolysin showed the highest inhibition of β-carotene oxidation, while the hydrolysate obtained with Alcalase and thermolysin yielded the highest inhibition of low-density lipoprotein peroxidation [66]. The same authors also employed an assay based on the protection of DNA molecules against oxidative damage. This assay evaluated the capacity of peptides to avoid the scission of supercoiled DNA strands in the presence of peroxyl and hydroxyl radicals. Peptides obtained by the hydrolysis of date seed proteins with Alcalase showed the highest capacity to inhibit DNA oxidative damage. Choudhary et al. [30] employed another assay that was also based on the inhibition of the oxidative degradation of DNA to study antioxidant activity of proteins from bottle gourd seeds. In this case, Cu^{2+} and H_2O_2 were employed to induce DNA oxidation.

All these methods are based on in vitro reactions, and though they are very useful as first screenings for potential antioxidants, further studies are required to confirm the real antioxidant capacity of peptides and proteins. Methods using cell cultures or those measuring antioxidant molecules in plasma or tissues from in vivo assays have been employed to confirm in vitro results.

Antioxidant activity has been evaluated by the determination of the level of intracellular ROS produced when cells are submitted to oxidative stress by the addition of a peroxide (hydrogen peroxide or tertbutylhydroperoxide). For that purpose, a fluorescence probe (2′, 7′-dichloro-dihydro-fluorescein diacetate) was employed. This molecule is hydrolyzed by intracellular esterase to form 2′, 7′-dichloro-dihydro-fluorescein. These molecules are next oxidized by intracellular ROS, which are produced under oxidative stress, and result in a highly fluorescent molecule (2′, 7′-dichloro-fluorescein) that can be monitored by fluorescence spectroscopy. This assay was employed to compare antioxidant capacity of peptides obtained from different genotypes of *Prunus* fruits and olive seeds. The measurement of ROS produced by cervical cancer cells (HeLa cells) under oxidizing conditions in the presence or absence of hydrolysates confirmed the antioxidant capacity of peptides, although no significant differences were observed among genotypes. This assay was also employed to measure antioxidant activity in five synthetic peptides found in watermelon seeds. The presence of a peptide with the Arg–Asp–Pro–Glu–Glu–Arg sequence reduced the generation of ROS in HepG2 cells under oxidizing conditions (H_2O_2).

Furthermore, ROS can increase cell membrane permeability, thus resulting in high intracellular concentrations of Ca^{2+}. Intracellular Ca^{2+} concentration can be measured by fluorescence spectroscopy after adding a fluorescent dye. This assay demonstrated that the presence of peptide Arg–Asp–Pro–Glu–Glu–Arg reduced intracellular Ca^{2+} concentration in H_2O_2-damaged HepG2 cells and, thus, membrane cell damages generated under oxidizing conditions. Furthermore, an additional assay involving the use of two different DNA dyes has been employed to determine cell membrane oxidative damages. The acridine orange dye can penetrate into normal cell membranes and stain DNA into green, while ethidium bromide dye can penetrate into damaged cell membranes to stain DNA into orange or red. HepG2 cells under H_2O_2-induced-oxidizing stress showed less red cells when the Arg–Asp–Pro–Glu–Glu–Arg peptide was present [51].

Moreover, cancer cell initiation and progression has been linked to oxidative stress. In order to demonstrate that potential antioxidant peptides can decrease the proliferation of cancer cells and exert protective effects, cell viability can be determined using the MTT (3-(4,5-dimethylthiazol-2-yl)-2,5-diphenyltetrazolium bromide) assay. This assay measures the metabolic activity of cells through oxidation–reduction reactions happening in mitochondria via a succinate dehydrogenase system. The reduction of MTT in the mitochondria results in blue insoluble formazan that is measured by spectroscopy [16]. Studies on cell viability have demonstrated the lack of a cytotoxic effect of *Prunus* fruits and olive seed hydrolysates in normal HK-2 cells and their antiproliferative effect in malignant cells from human prostate cancer, colorectal adenocarcinoma, and cervical cancer. The MTT assay was also employed to demonstrate the cytoprotective effect of watermelon seed peptides with molecular weights below 1 kDa on RAW 264.7 cells submitted to oxidative stress induced by H_2O_2. The same authors also measured the nuclear factor erythroid 2-related factor 2 (Nrf2) and heme oxygenase-1 (HO-1) proteins, which participate in one of the most important antioxidant pathways in cells after treatment with H_2O_2. Nrf2 and HO-1 proteins were more expressed in cells when WSPHs-1 peptides from watermelon seeds were present [53].

The antioxidant activity of proteins has also been studied in some fruit residues. Different authors compared the antioxidant activity of fractions obtained by the application of the Osborne method to the extraction of proteins from bottle gourd [31], watermelon [49], pumpkin (*Cucurbita moschata*) [20], and melon [47] seeds. The fraction exhibiting the highest antioxidant activity in the case of the bottle gourd, watermelon, and pumpkin seeds was that containing globulins, while the glutelin fraction showed the lowest activity. As expected, proteins in globulin fractions were found to hold the highest amount of hydrophobic amino acids. The opposite behavior was observed for melon seed proteins. In this case, the globulin fraction yielded the least antioxidant activity, while the glutelin fraction showed the most and was also the one with the highest presence of polyphenols.

Previous studies have demonstrated that protein malnourishment lead to overoxidation [57]. Under these conditions, in vivo antioxidant system, consisting mainly of enzymes such as catalase, superoxide dismutase, glutathione peroxidase, and glutathione reductase, is not enough to avoid oxidative damage. The addition of antioxidants can increase the activity of these antioxidant enzymes. The antioxidant properties of proteins extracted from pumpkin seeds were investigated on low-protein fed rats submitted to CCl_4 intoxication. The authors found that feeding of rats with pumpkin seed proteins resulted in increasing activity levels of catalase, superoxide dismutase, and glutathione peroxidase in the rat plasma and in a decreasing lipid peroxidation [57]. Nevertheless, pumpkin seed proteins could not inhibit the activity of the xanthine oxidase enzyme that promotes the generation of free radicals [61]. Moreover, different *Cucurbitaceae* (watermelon, bottle gourd, and pumpkin (*Cucurbita moschata*)) seeds yielded protein hydrolysates that were able to increase catalase levels in normal mice while also reducing in vivo lipid peroxidation [32]. Additionally, the synthetic peptide Arg–Asp–Pro–Glu–Glu–Arg, found in watermelon protein hydrolysates, was found to increase superoxide dismutase, glutathione peroxidase, and catalase activity in HepG2 cells under H_2O_2 induced oxidative damage [51].

7. Peptide Fractionation

In some cases, hydrolysates have been fractionated to obtain highly active fractions. The main easier fractionation technique is ultrafiltration. Ultrafiltration is a membrane separation technique that uses cut-off filters as molecular sieves (usually from 3 to 10 kDa); thus, peptides can be separated according to their molecular size. The main disadvantage of this technique is its low selectivity.

The use of ultrafiltration has been found to not provide a fraction with higher antioxidant peptides than the parent hydrolysate in the case of the olive, peach, and cherry seeds [17,28,33]. This is justified when considering that antioxidants work in a collaborative way, and, thus, the deficiency of a component in an antioxidant system can affect the efficiency of other. This behavior is common among antioxidant compounds. Nevertheless, in Chinese cherries and tomato seeds, ultrafiltration has provided fractions that exhibited the greatest antioxidant activity [21,55]. In most cases, most active peptides were in fractions under 3 and/or 1 kDa [17,21,35,53,55,93]. However, peptides above 5 kDa from peach and cherry seeds showed similar antioxidant activities to peptides below 3 kDa, with the peptides between 3 and 5 kDa being less effective [28,33]. The low selectivity of ultrafiltration filters may be related to this result [33].

Size-exclusion chromatography (SEC), dialysis, and preparative reversed-phase HPLC (RP-HPLC) are other techniques employed for the fractionation of hydrolysates. SEC is a low resolution technique normally followed by an orthogonal fractionation using semipreparative RP-HPLC. Peptides from Chinese cherry, tomato, and *Benincasa hispida* seeds have been fractionated using SEC with the dextran resins (Sephadex) and mobile phases consisting of a phosphate buffer (pH 6.5) [44], hydrochloric acid [39], or distilled water [55].

C18-bound phases have been always employed for the fractionation of peptides by RP-HPLC using elution gradients and mobile phases consisting of water with trifluoroacetic acid (TFA) (mobile phase A) and acetonitrile (ACN) with TFA (mobile phase B). TFA acts as ion-pairing reagent to maintain a low pH, create complexes with positively charged peptides, and minimize their ionic interactions with the hydrophobic stationary phase [94].

In many occasions, the combination of different fractionation techniques enables the isolation of most active peptides. Peptides from apricot seeds [27] were fractionated by dialysis followed by SEC in Sephadex G-25 and G-15, as well as RP-HPLC. Hydrolysates from Chinese cherry [55], watermelon [51], tomato [39], and wax gourd (*Benincasa hispida*) [44] protein seeds were fractionated by combining ultrafiltration, SEC, and RP-HPLC.

8. Peptide Identification

Antioxidant peptides obtained by extraction and proteolysis have been sequenced using mass spectrometry (MS). Usually, parent proteins have not been previously sequenced and they are not available in databases. In these cases, peptides are identified by de novo sequencing by the direct analysis of tandem mass spectra [3,14,16,17,25,33,35]. In a few cases, proteins from fruit wastes are present in databases, and a strategy based on database searching enables the identification of peptide sequence [21,29].

The identification of antioxidant peptides has enabled the observation of some common features within them, such as a high amount of hydrophobic (leucine/isoleucine, alanine, methionine, threonine, glycine, valine, and proline) [3,14,16,25,27,28,39,42,55] and aromatic amino acids (phenylalanine, tyrosine, tryptophan, and histidine) [3,14,16,21,25,27,28,39,40,47,55], as well as molecular weights below 1 kDa [14,16,25,27,28].

MS can directly provide information on the mass of a particular peptide but can also generate amino acid sequence information from tandem mass spectra (MS/MS). The MS analysis of peptides is possible using electrospray ionization (ESI) and matrix-assisted laser desorption/ionization (MALDI). Soft ionization allows for the transfer of polypeptide ions into the gas phase without their insource fragmentation. The identification of peptides by MS is carried out in the positive ion mode. In most cases, high resolution quadrupole time-of-flight (Q-TOF) or TOF mass analyzers are employed.

A MALDI source results mainly in singly charged ions and is considered a robust method of ionization in the presence of salts and detergents, much less prone to ionization suppression effects than ESI [94]. However, MALDI requires off-line sample deposition onto a target plate, and it is less convenient to couple with HPLC. Two peptides with molecular masses of 673.1 Da (Val–Leu–Tyr–Ile–Trp) and 566.9 Da (Ser–Val–Pro–Tyr–Glu) were identified in the most antioxidant fraction obtained from apricot seeds [27]. In other works with jujube and tomato seeds, most antioxidant peptides were observed in the mass range from 7 to 16 kDa in the case of jujube seeds [45] and from 0.5 to 0.8 Da and 1.2 to 1.5 Da in the case of tomato seeds [40], according to data obtained by MALDI-TOF. On the other hand, the molecular mass of "hispidalin" peptide from wax gourd seeds was 5.7 kDa, with the products of its hydrolysis being between 1.0 and 1.8 kDa [44].

Guo et al. [55] identified two peptides (Phe–Pro–Glu–Leu–Leu–Ile and Val–Phe–Ala–Ala–Leu) as main contributors to the antioxidant activity in Chinese cherry seed hydrolysate using just a Q-TOF MS and no coupling to a chromatographic separation. However, the analysis of hydrolysates by ESI-MS normally requires and is carried out by the previous separation of hydrolysates by HPLC. RP-HPLC is the most common chromatographic mode due to its high efficiency and the compatibility of mobile phases with ESI desorption. In this case, TFA, widely used as a counterion in the separation of peptides by RP-HPLC, is replaced by acetic acid or formic acid since TFA results in strong signal suppression in MS [3,16,17,33].

Hydrophilic interaction chromatography (HILIC) may also be useful for the separation of highly polar compounds that cannot be retained on RP-HPLC. The separation of analytes by HILIC is based on the interaction with a hydrophilic stationary phase like in normal-phase chromatography. However, HILIC uses water-miscible solvents (e.g., ACN), and elution is achieved by a water gradient that makes this technique suitable for coupling with MS. Both RP-HPLC and HILIC have been employed in the identification of peptides in plum, cherry, and peach seeds [28,33,35] and in pomegranate peel [25]. Figure 2 shows the total ion chromatogram obtained by RP-HPLC-ESI-QTOF and HILIC-ESI-Q-TOF of a hydrolysate obtained from plum seed proteins using Alcalase. Additionally, Figure 2 shows the mass spectrum of peptide His–Leu–Pro–Pro–Leu–Leu that was observed in both chromatographic modes. A comparison of peptides identified in plum, peach, and cherry seeds enabled the observation of some common sequences within *Prunus* fruits: Leu–Tyr–Ser–Pro–His, Leu–Tyr–Thr–Pro–His, Leu–Leu–Ala–Gln–Ala, Leu–Ala–Gly–Asn–Pro–Glu–Asn–Glu, Leu–Leu–Asn–Asp–Glu, and Leu–Leu–Met–Gln. The use of both chromatographic modes enabled, in all cases, a more comprehensive identification of peptides. Peptides identified using both chromatographic modes in hydrolysates obtained from pomegranate peel proteins extracted using HIFU, PLE, and DES were compared by Hernández-Corroto et al. [3,25]. Despite there being common peptides in the three hydrolysates, there were also many different peptides within them. This was attributed to the extraction of different kind of proteins by every technique [3,25].

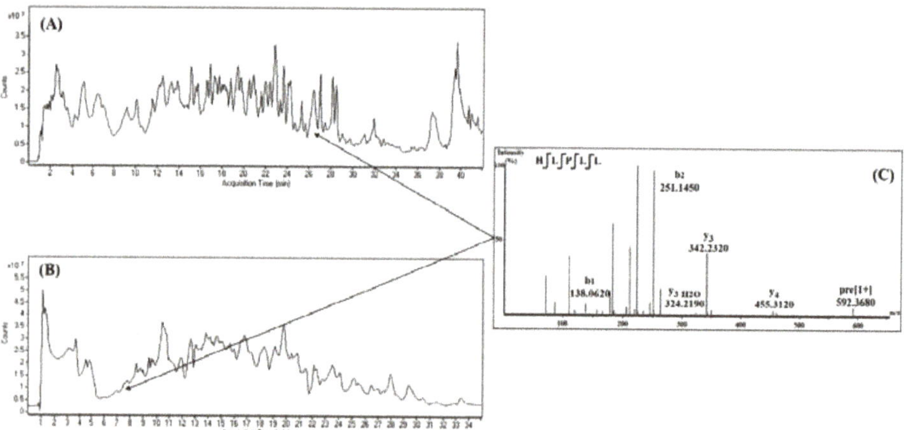

Figure 2. Total ion chromatograms corresponding to a hydrolysate obtained from plum seed proteins using Alcalase hydrolysate obtained by RP-HPLC (**A**), HILIC (**B**), and MS/MS spectrum of a peptide simultaneously observed in both chromatographic modes (**C**). With permission from [35].

Wen et al. employed a quadrupole-orbitrap MS instrument for the identification of most potent peptides in watermelon seeds. Five peptides were identified and then synthesized for their further characterization: Arg–Asp–Pro–Glu–Glu–Arg, Lys–Glu–Leu–Glu–Glu–Lys, Asp–Ala–Ala–Gly–Arg–Leu–Gln–Glu, Leu–Asp–Asp–Asp–Gly–Arg–Leu, and Gly–Phe–Ala–Gly–Asp–Asp–Ala–Pro–Arg–Ala [51]. The peptide Arg–Asp–Pro–Glu–Glu–Arg showed the highest antioxidant activity, according to results obtained by ABTS, DPPH, and oxygen radical antioxidant capacity (ORAC) assays. Moreover, it also showed cytoprotective effects on HepG2 cells under induced oxidative stress.

The introduction of nanoelectrospray (nanoESI) has enabled an increase in sensitivity. Peptides in most antioxidant fractions from fermented tomato seeds were identified by nLC/MS/MS (QTOF), obtaining many sequences below 600 Da and 5–6 amino acid residues. Among peptides, GQVPP showed a significant antioxidant activity (97% DPPH scavenging activity at 0.4 mM) [39].

None of the bioactive peptides cited in this section of the review were previously recorded in the BIOPEP database.

9. Conclusions

Antioxidant peptides have been discovered in some fruit residues (largely seeds and peels). Most of them have been isolated by solid–liquid conventional extraction under basic conditions. Ultrasound-assisted extraction, pressurized liquid extraction, microwave-assisted extraction, and the application of electrical energy (pulsed electric fields or high-voltage electrical discharges) have been employed in order to improve the sustainability and yield of extraction. An increasing trend is the use of non-polluting solvents such as deep eutectic ones. Future progress in this area will likely be focused in the combination of different (orthogonal) extraction techniques and strategies. Some works have already demonstrated that when using a combination of DES with HIFU or MAE, orpretreatment with HVED or PEF. The extraction of protein accomplishes higher yields—however, further research in this direction is required to attain solid developments in this field.

While non-conventional methods are promising because of their many advantages (such as the reduction of processing time, the use of greener and/or safer solvents, and higher extraction yields), most of them are not ready for scaling up in the industry. Extraction and purification with carbosilane dendrimers is another worthwhile proposal, although it has not been studied much yet.

Some antioxidant peptides were directly extracted from these residues, but most were found in a latent state as part of protein sequences. The release of these peptides was mainly achieved by enzymatic digestion, with Alcalase being the enzyme that usually resulted in higher antioxidant peptides. The use of unrelated in vitro assays based on different antioxidant mechanisms guarantees a comprehensive evaluation of a fruit residue's peptide capacity. The evaluation of the capacity of peptides to scavenge free radicals, inhibit lipid peroxidation, or reduce oxidants is the most popular study area, although further experiments using cell cultures and animal models are essential to confirm in vitro results. The fractionation of peptides by ultrafiltration, size-exclusion chromatography, and reversed-phase chromatography has proven useful to isolate most antioxidant peptides. Tandem mass spectrometry has enabled the identification of peptide sequences with significant antioxidant properties that, in some cases, have been synthesized for further study. In general, antioxidant peptides from fruit seeds have shown short sequences and contain hydrophobic and aromatic amino acids.

The research discussed in this review shows that fruit residues store great amounts of antioxidant peptides, which are highly valuable products with promising applications as nutraceuticals or added to functional foods. As such, harvesting these biomolecules may represent a partial solution to the increasing environmental concerns about the management of fruit residues.

Author Contributions: S.O.-G. and M.C.G. wrote the paper. M.C.G. and M.L.M. supervised the project. All authors provided critical feedback and helped shape the research, analysis, and manuscript. All authors have read and agreed to the published version of the manuscript.

Funding: This work was supported by the Spanish Ministry of Economy and Competitiveness (ref. AGL2016-79010-R) and the Comunidad de Madrid (Spain) and European funding from FSE and FEDER programs (S2018/BAA-4393, AVANSECAL-II-CM).

Acknowledgments: S. Olivares thanks the Community of Madrid for his contract as research assistant.

Conflicts of Interest: The authors declare no conflict of interest.

References

1. United Nations, Department of Economic and Social Affairs, Population Division. *World Population Prospects 2019; 1: Comprensive Tables, 2–4*; United Nations: New York, NY, USA, 2019.
2. Global Production of Fresh Fruit from 1990 to 2018 (in Million Metric Tons). Available online: https://www.statista.com/statistics/262266/global-production-of-fresh-fruit/ (accessed on 16 April 2020).
3. Hernández-Corroto, E.; Marina, M.L.; García, M.C. Extraction and identification by high resolution mass spectrometry of bioactive substances in different extracts obtained from pomegranate peel. *J. Chromatogr. A* **2019**, *1594*, 82–92. [CrossRef] [PubMed]
4. *Communication from the Commision to the Council and the European Parliament on the Interpretative Communication on Waste and By-Products*; Commision of the European Communities: Brussels, Belguim, 2007.
5. Padayachee, A.; Day, L.; Howell, K.; Gidley, M. Complexity and Health Functionality of Plant Cell Wall Fibres from Fruits and Vegetables. *Crit. Rev. Food Sci. Nutr.* **2015**, *57*, 59–81. [CrossRef] [PubMed]
6. Dhillon, G.; Dhillon, S.; Brar, S. Perspective of apple processing wastes as low-cost substrates for bioproduction of high value products: A review. *Renew. Sustain. Energy Rev.* **2013**, *27*, 789–805. [CrossRef]
7. Wang, L.; Shen, F.; Yuan, H.; Zou, D.; Liu, Y.; Zhu, B.; Li, X. Anaerobic co-digestion of kitchen waste and fruit/vegetable waste: Lab-scale and pilot-scale studies. *Waste Manag.* **2014**, *34*, 2627–2633. [CrossRef]
8. Burange, A.; Clark, J.H.; Luque, R. Trends in Food and Agricultural Waste Valorization. In *Encyclopedia of Inorganic and Bioinorganic Chemistry*; John Wiley & Sons, Ltd.: Hoboken, NJ, USA, 2016; pp. 1–10, ISBN 978-1-119-95143-8.
9. Deng, G.-F.; Shen, C.; Xu, X.-R.; Kuang, R.-D.; Guo, Y.-J.; Zeng, L.-S.; Gao, L.-L.; Lin, X.; Xie, J.-F.; Xia, E.-Q.; et al. Potential of fruit wastes as natural resources of bioactive compounds. *Int. J. Mol. Sci.* **2012**, *13*, 8308–8323. [CrossRef]
10. Hajfathalian, M.; Ghelichi, S.; García-Moreno, P.J.; Moltke Sørensen, A.-D.; Jacobsen, C. Peptides: Production, bioactivity, functionality, and applications. *Crit. Rev. Food Sci. Nutr.* **2018**, *58*, 3097–3129. [CrossRef]
11. Garcia, M.C.; Gonzalez-Garcia, E.; Vasquez-Villanueva, R.; Marina, M.L. Apricot and other seed stones: Amygdalin content and the potential to obtain antioxidant, angiotensin I converting enzyme inhibitor and hypocholesterolemic peptides. *Food Funct.* **2016**, *7*, 4693–4701. [CrossRef]
12. García, M.C.; Puchalska, P.; Esteve, C.; Marina, M.L. Vegetable foods: A cheap source of proteins and peptides with antihypertensive, antioxidant, and other less occurrence bioactivities. *Talanta* **2013**, *106*, 328–349. [CrossRef]
13. Gonzalez-Garcia, E.; Garcia, M.C.; Marina, M.L. Capillary liquid chromatography-ion trap-mass spectrometry methodology for the simultaneous quantification of four angiotensin-converting enzyme-inhibitory peptides in Prunus seed hydrolysates. *J. Chromatogr. A* **2018**, *1540*, 47–54. [CrossRef]
14. González-García, E.; Marina, M.L.; García, M.C. Plum (Prunus Domestica L.) by-product as a new and cheap source of bioactive peptides: Extraction method and peptides characterization. *J. Funct. Foods* **2014**, *11*, 428–437. [CrossRef]
15. Minkiewicz, P.; Iwaniak, A.; Darewicz, M. BIOPEP-UWM database of bioactive peptides: Current opportunities. *Int. J. Mol. Sci.* **2019**, *20*, 5978. [CrossRef] [PubMed]
16. Hernandez-Corroto, E.; Marina, M.L.; Garcia, M.C. Multiple protective effect of peptides released from Olea europaea and Prunus persica seeds against oxidative damage and cancer cell proliferation. *Food Res. Int.* **2018**, *106*, 458–467. [CrossRef] [PubMed]
17. Esteve, C.; Marina, M.L.; Garcia, M.C. Novel strategy for the revalorization of olive (Olea europaea) residues based on the extraction of bioactive peptides. *Food Chem.* **2015**, *167*, 272–280. [CrossRef] [PubMed]
18. Aluko, R.E. 5-Amino acids, peptides, and proteins as antioxidants for food preservation. In *Handbook of Antioxidants for Food Preservation*; Shahidi, F., Ed.; Woodhead Publishing: Cambridge, UK, 2015; pp. 105–140, ISBN 978-1-78242-089-7.
19. Samaranayaka, A.G.P.; Li-Chan, E.C.Y. Food-derived peptidic antioxidants: A review of their production, assessment, and potential applications. *J. Funct. Foods* **2011**, *3*, 229–254. [CrossRef]
20. Dash, P.; Ghosh, G. Proteolytic and antioxidant activity of protein fractions of seeds of Cucurbita moschata. *Food Biosci.* **2017**, *18*, 1–8. [CrossRef]

21. Meshginfar, N.; Sadeghi Mahoonak, A.; Hosseinian, F.; Ghorbani, M.; Tsopmo, A. Production of antioxidant peptide fractions from a by-product of tomato processing: Mass spectrometry identification of peptides and stability to gastrointestinal digestion. *J. Food Sci. Technol.* **2018**, *55*, 3498–3507. [CrossRef]

22. Karami, Z.; Akbari-Adergani, B. Bioactive food derived peptides: A review on correlation between structure of bioactive peptides and their functional properties. *J. Food Sci. Technol.* **2019**, *56*, 535–547. [CrossRef]

23. Mäkinen, S.; Johannson, T.; Vegarud Gerd, E.; Pihlava, J.M.; Pihlanto, A. Angiotensin I-converting enzyme inhibitory and antioxidant properties of rapeseed hydrolysates. *J. Funct. Foods* **2012**, *4*, 575–583. [CrossRef]

24. Barba, F.; Boussetta, N.; Vorobiev, E. Emerging Technologies for the Recovery of Isothiocyanates, Protein and Phenolic Compounds from Rapeseed and Rapeseed Press-Cake: Effect of High Voltage Electrical Discharges. *Innov. Food Sci. Emerg. Technol.* **2015**, *31*, 67–72. [CrossRef]

25. Hernández-Corroto, E.; Plaza, M.; Marina, M.L.; García, M.C. Sustainable extraction of proteins and bioactive substances from pomegranate peel (Punica granatum L.) using pressurized liquids and deep eutectic solvents. *Innov. Food Sci. Emerg. Technol.* **2020**, *60*, 102314. [CrossRef]

26. Parniakov, O.; Barba, F.J.; Grimi, N.; Lebovka, N.; Vorobiev, E. Extraction assisted by pulsed electric energy as a potential tool for green and sustainable recovery of nutritionally valuable compounds from mango peels. *Food Chem.* **2016**, *192*, 842–848. [CrossRef] [PubMed]

27. Zhang, H.; Xue, J.; Zhao, H.; Zhao, X.; Xue, H.; Sun, Y.; Xue, W. Isolation and Structural Characterization of Antioxidant Peptides from Degreased Apricot Seed Kernels. *J. AOAC Int.* **2019**, *101*, 1661–1663. [CrossRef] [PubMed]

28. Vasquez-Villanueva, R.; Marina, M.L.; Garcia, M.C. Identification by hydrophilic interaction and reversed-phase liquid chromatography-tandem mass spectrometry of peptides with antioxidant capacity in food residues. *J. Chromatogr. A* **2016**, *1428*, 185–192. [CrossRef] [PubMed]

29. Gonzalez-Garcia, E.; Marina, M.L.; Garcia, M.C.; Righetti, P.G.; Fasoli, E. Identification of plum and peach seed proteins by nLC-MS/MS via combinatorial peptide ligand libraries. *J. Proteom.* **2016**, *148*, 105–112. [CrossRef]

30. Choudhary, R.R.; Verma, H.N.; Sharma, N.K. Antioxidant and Dehalogenase Activity of Lagenaria sciceraria Seed Protein Fraction. *J. Plant. Sci. Res.* **2013**, *29*, 145–158.

31. Dash, P.; Panda, P.; Ghosh, G. Free Radical Scavenging Activities and Nutritional Value of Lagenaria siceraria: A Nutriment Creeper. *Iran. J. Sci. Technol. Trans. A Sci.* **2018**, *42*, 1743–1752. [CrossRef]

32. Dash, P.; Rath, G.; Ghosh, G. In Vivo Antioxidant Potential of Protein Hydrolysates of some Cucurbitaceae Seed. *JDDT [Internet]* **2020**, *10*, 128–132.

33. García, M.C.; Endermann, J.; González-García, E.; Marina, M.L. HPLC-Q-TOF-MS Identification of Antioxidant and Antihypertensive Peptides Recovered from Cherry (Prunus cerasus L.) Subproducts. *J. Agric. Food Chem.* **2015**, *63*, 1514–1520. [CrossRef]

34. Rosélló-Soto, E.; Barba, F.J.; Parniakov, O.; Galanakis, C.M.; Lebovka, N.; Grimi, N.; Vorobiev, E. High Voltage Electrical Discharges, Pulsed Electric Field, and Ultrasound Assisted Extraction of Protein and Phenolic Compounds from Olive Kernel. *Food Bioprocess. Technol.* **2015**, *8*, 885–894. [CrossRef]

35. González-García, E.; Puchalska, P.; Marina, M.L.; García, M.C. Fractionation and identification of antioxidant and angiotensin-converting enzyme-inhibitory peptides obtained from plum (Prunus domestica L.) stones. *J. Funct. Foods* **2015**, *19*, 376 384. [CrossRef]

36. Mechmeche, M.; Kachouri, F.; Ksontini, H.; Hamdi, M. Production of bioactive peptides from tomato seed isolate by Lactobacillus plantarum fermentation and enhancement of antioxidant activity. *Food Biotechnol.* **2017**, *31*, 94–113. [CrossRef]

37. Mechmeche, M.; Ksontini, H.; Hamdi, M.; Kachouri, F. Production of Bioactive Peptides in Tomato Seed Protein Isolate Fermented by Water Kefir Culture: Optimization of the Fermentation Conditions. *Int. J. Pept. Res. Ther.* **2019**, *25*, 137–150. [CrossRef]

38. Moayedi, A.; Hashemi, M.; Safari, M. Valorization of tomato waste proteins through production of antioxidant and antibacterial hydrolysates by proteolytic Bacillus subtilis: Optimization of fermentation conditions. *J. Food Sci. Technol.* **2016**, *53*, 391–400. [CrossRef] [PubMed]

39. Moayedi, A.; Mora, L.; Aristoy, M.C.; Safari, M.; Hashemi, M.; Toldrá, F. Peptidomic analysis of antioxidant and ACE-inhibitory peptides obtained from tomato waste proteins fermented using Bacillus subtilis. *Food Chem.* **2018**, *250*, 180–187. [CrossRef] [PubMed]

40. Moayedi, A.; Mora, L.; Aristoy, M.-C.; Hashemi, M.; Safari, M.; Toldrá, F. ACE-Inhibitory and Antioxidant Activities of Peptide Fragments Obtained from Tomato Processing By-Products Fermented Using Bacillus subtilis: Effect of Amino Acid Composition and Peptides Molecular Mass Distribution. *Appl. Biochem. Biotechnol.* **2017**, *181*, 48–64. [CrossRef]

41. Mechmeche, M.; Kachouri, F.; Ksontini, H.; Setti, K.; Hamdi, M. Bioprocess development and preservation of functional food from tomato seed isolate fermented by kefir culture mixture. *J. Food Sci. Technol.* **2018**, *55*, 3911–3921. [CrossRef]

42. Mechmeche, M.; Kachouri, F.; Chouabi, M.; Ksontini, H.; Setti, K.; Hamdi, M. Optimization of Extraction Parameters of Protein Isolate from Tomato Seed Using Response Surface Methodology. *Food Anal. Methods* **2017**, *10*, 809–819. [CrossRef]

43. Meshginfar, N.; Mahoonak, A.S.; Hosseinian, F.; Tsopmo, A. Physicochemical, antioxidant, calcium binding, and angiotensin converting enzyme inhibitory properties of hydrolyzed tomato seed proteins. *J. Food Biochem.* **2019**, *43*, e12721. [CrossRef]

44. Sharma, S.; Verma, H.N.; Sharma, N.K. Cationic Bioactive Peptide from the Seeds of Benincasa hispida. *Int. J. Pept.* **2014**, *2014*, 156060. [CrossRef]

45. Kanbargi, K.D.; Sonawane, S.K.; Arya, S.S. Functional and antioxidant activity of Ziziphus jujube seed protein hydrolysates. *J. Food Meas. Charact.* **2016**, *10*, 226–235. [CrossRef]

46. Kanbargi, K.D.; Sonawane, S.K.; Arya, S.S. Encapsulation characteristics of protein hydrolysate extracted from Ziziphus jujube seed. *Int. J. Food Prop.* **2017**, *20*, 3215–3224. [CrossRef]

47. Siddeeg, A.; Xu, Y.; Jiang, Q.; Xia, W. In vitro antioxidant activity of protein fractions extracted from seinat (Cucumis melo var. tibish) seeds. *CYTA J. Food* **2015**, *13*, 1–10. [CrossRef]

48. Arise, R.; Yekeen, A.; Ekun, O.; Olatomiwa, O. Protein Hydrolysates from Citrullus lanatus Seed: Antiradical and Hydrogen Peroxide-scavenging properties and kinetics of Angiotensin-I converting enzyme inhibition. *Ceylon J. Sci.* **2016**, *45*, 39. [CrossRef]

49. Dash, P.; Ghosh, G. Fractionation, amino acid profiles, antimicrobial and free radical scavenging activities of Citrullus lanatus seed protein. *Nat. Prod. Res.* **2017**, *31*, 2945–2947. [CrossRef] [PubMed]

50. Arise, R.; Yekeen, A.; Ekun, O. In vitro antioxidant and α-amylase inhibitory properties of watermelon seed protein hydrolysates. *Environ. Exp. Biol.* **2016**, *14*, 163–172. [CrossRef]

51. Wen, C.; Zhang, J.; Feng, Y.; Duan, Y.; Ma, H.; Zhang, H. Purification and identification of novel antioxidant peptides from watermelon seed protein hydrolysates and their cytoprotective effects on H2O2-induced oxidative stress. *Food Chem.* **2020**, *327*, 127059. [CrossRef] [PubMed]

52. Sonawane, S.K.; Arya, S.S. Citrullus lanatus protein hydrolysate optimization for antioxidant potential. *J. Food Meas. Charact.* **2017**, *11*, 1834–1843. [CrossRef]

53. Wen, C.; Zhang, J.; Zhang, H.; Duan, Y.; Ma, H. Effects of divergent ultrasound pretreatment on the structure of watermelon seed protein and the antioxidant activity of its hydrolysates. *Food Chem.* **2019**, *299*, 125165. [CrossRef]

54. Parniakov, O.; Roselló-Soto, E.; Barba, F.J.; Grimi, N.; Lebovka, N.; Vorobiev, E. New approaches for the effective valorization of papaya seeds: Extraction of proteins, phenolic compounds, carbohydrates, and isothiocyanates assisted by pulsed electric energy. *Food Res. Int.* **2015**, *77*, 711–717. [CrossRef]

55. Guo, P.; Qi, Y.; Zhu, C.; Wang, Q. Purification and identification of antioxidant peptides from Chinese cherry (Prunus pseudocerasus Lindl.) seeds. *J. Funct. Foods* **2015**, *19*, 394–403. [CrossRef]

56. Osukoya, O.; Nwoye-Ossy, M.; Olayide, I.; Ojo, O.; Adewale, O.; Kuku, A. Antioxidant activities of peptide hydrolysates obtained from the seeds of Treculia africana Decne (African breadfruit). *Prep. Biochem. Biotechnol.* **2020**, *50*, 1–7. [CrossRef] [PubMed]

57. Nkosi, C.Z.; Opoku, A.R.; Terblanche, S.E. Antioxidative effects of pumpkin seed (Cucurbita pepo) protein isolate in CCl4-induced liver injury in low-protein fed rats. *Phytother. Res.* **2006**, *20*, 935–940. [CrossRef] [PubMed]

58. Vaštag, Ž.; Popović, L.; Popović, S.; Krimer, V.; Peričin, D. Production of enzymatic hydrolysates with antioxidant and angiotensin-I converting enzyme inhibitory activity from pumpkin oil cake protein isolate. *Food Chem.* **2011**, *124*, 1316–1321. [CrossRef]

59. Nourmohammadi, E.; SadeghiMahoonak, A.; Alami, M.; Ghorbani, M. Amino acid composition and antioxidative properties of hydrolysed pumpkin (Cucurbita pepo L.) oil cake protein. *Int. J. Food Prop.* **2017**, *20*, 3244–3255. [CrossRef]

60. Fan, S.; Hu, Y.; Li, C.; Liu, Y. Optimization of preparation of antioxidative peptides from pumpkin seeds using response surface method. *PLoS ONE* **2014**, *9*, e92335. [CrossRef]

61. Nkosi, C.Z.; Opoku, A.R.; Terblanche, S.E. In Vitro antioxidative activity of pumpkin seed (Cucurbita pepo) protein isolate and its In Vivo effect on alanine transaminase and aspartate transaminase in acetaminophen-induced liver injury in low protein fed rats. *Phytother. Res.* **2006**, *20*, 780–783. [CrossRef] [PubMed]

62. Popović, L.; Peričin, D.; Vaštag, Ž.; Popović, S.; Krimer, V.; Torbica, A. Antioxidative and Functional Properties of Pumpkin Oil Cake Globulin Hydrolysates. *J. Am. Oil Chem. Soc.* **2013**, *90*, 1157–1165. [CrossRef]

63. Venuste, M.; Zhang, X.; Shoemaker, C.F.; Karangwa, E.; Abbas, S.; Kamdem, P.E. Influence of enzymatic hydrolysis and enzyme type on the nutritional and antioxidant properties of pumpkin meal hydrolysates. *Food Funct.* **2013**, *4*, 811–820. [CrossRef]

64. Liu, R.-L.; Yu, P.; Ge, X.-L.; Bai, X.-F.; Li, X.-Q.; Fu, Q. Establishment of an Aqueous PEG 200-Based Deep Eutectic Solvent Extraction and Enrichment Method for Pumpkin (Cucurbita moschata) Seed Protein. *Food Anal. Methods* **2017**, *10*, 1669–1680. [CrossRef]

65. Ambigaipalan, P.; Al-Khalifa, A.S.; Shahidi, F. Antioxidant and angiotensin I converting enzyme (ACE) inhibitory activities of date seed protein hydrolysates prepared using Alcalase, Flavourzyme and Thermolysin. *J. Funct. Foods* **2015**, *18*, 1125–1137. [CrossRef]

66. Ambigaipalan, P.; Shahidi, F. Antioxidant potential of date (Phoenix dactylifera L.) seed protein hydrolysates and carnosine in food and biological systems. *J. Agric. Food Chem.* **2015**, *63*, 864–871. [CrossRef] [PubMed]

67. Romelle, F.D.; Rani, A.; Manohar, R.S. Chemical composition of some selected fruit peels. *Eur. J. Food Sci. Technol.* **2016**, *4*, 12–21.

68. Lima, B.N.B.; Lima, F.F.; Tavares, M.I.B.; Costa, A.M.M.; Pierucci, A.P.T.R. Determination of the centesimal composition and characterization of flours from fruit seeds. *Food Chem.* **2014**, *151*, 293–299. [CrossRef] [PubMed]

69. Raihana, A.R.N.; Marikkar, J.M.N.; Amin, I.; Shuhaimi, M. A Review on Food Values of Selected Tropical Fruits' Seeds. *Int. J. Food Prop.* **2015**, *18*, 2380–2392. [CrossRef]

70. Lam, S.K.; Ng, T.B. Passiflin, a novel dimeric antifungal protein from seeds of the passion fruit. *Phytomed. Int. J. Phytother. Phytopharm.* **2009**, *16*, 172–180. [CrossRef]

71. Osborne, T.B. *The Vegetable Proteins*, 2nd ed.; Longmans Green and Co: London, UK, 1924.

72. Ozuna, C.; Paniagua-Martínez, I.; Castaño-Tostado, E.; Ozimek, L.; Amaya-Llano, S.L. Innovative applications of high-intensity ultrasound in the development of functional food ingredients: Production of protein hydrolysates and bioactive peptides. *Food Res. Int.* **2015**, *77*, 685–696. [CrossRef]

73. Belwal, T.; Ezzat, S.M.; Rastrelli, L.; Bhatt, I.D.; Daglia, M.; Baldi, A.; Devkota, H.P.; Orhan, I.E.; Patra, J.K.; Das, G.; et al. A critical analysis of extraction techniques used for botanicals: Trends, priorities, industrial uses and optimization strategies. *TRAC Trends Anal. Chem.* **2018**, *100*, 82–102. [CrossRef]

74. González-García, E.; Maly, M.; de la Mata, F.J.; Gómez, R.; Marina, M.L.; García, M.C. Proof of concept of a "greener" protein purification/enrichment method based on carboxylate-terminated carbosilane dendrimer-protein interactions. *Anal. Bioanal. Chem.* **2016**, *408*, 7679–7687. [CrossRef]

75. González-García, E.; Sánchez-Nieves, J.; de la Mata, F.J.; Marina, M.L.; García, M.C. Feasibility of cationic carbosilane dendrimers for sustainable protein sample preparation. *Colloids Surf. B Biointerfaces* **2020**, *186*, 110746. [CrossRef]

76. González-García, E.; Gutiérrez Ulloa, C.E.; de la Mata, F.J.; Marina, M.L.; García, M.C. Sulfonate-terminated carbosilane dendron-coated nanotubes: A greener point of view in protein sample preparation. *Anal. Bioanal. Chem.* **2017**, *409*, 5337–5348. [CrossRef]

77. Vásquez-Villanueva, R.; Pena González, C.; Sánchez-Nieves, J.; Mata, F.; Marina, M.; García, M. Gold nanoparticles coated with carbosilane dendrons in protein sample preparation. *Microchim. Acta* **2019**, *186*, 508. [CrossRef] [PubMed]

78. Grosso, C.; Valentão, P.; Ferreres, F.; Andrade, P.B. Alternative and efficient extraction methods for marine-derived compounds. *Mar. Drugs* **2015**, *13*, 3182–3230. [CrossRef] [PubMed]

79. Mustafa, A.; Turner, C. Pressurized liquid extraction as a green approach in food and herbal plants extraction: A review. *Anal. Chim. Acta* **2011**, *703*, 8–18. [CrossRef] [PubMed]

80. Herrero, M.; Simó, C.; Ibáñez, E.; Cifuentes, A. Capillary electrophoresis-mass spectrometry of Spirulina platensis proteins obtained by pressurized liquid extraction. *Electrophor.* **2005**, *26*, 4215–4224. [CrossRef]

81. Abbott, A.P.; Capper, G.; Davies, D.L.; Rasheed, R.K.; Tambyrajah, V. Novel solvent properties of choline chloride/urea mixtures. *Chem. Commun.* **2003**, *1*, 70–71. [CrossRef]

82. Lores, H.; Romero, V.; Costas, I.; Bendicho, C.; Lavilla, I. Natural deep eutectic solvents in combination with ultrasonic energy as a green approach for solubilisation of proteins: Application to gluten determination by immunoassay. *Talanta* **2017**, *162*, 453–459. [CrossRef]

83. Paiva, A.; Craveiro, R.; Aroso, I.; Martins, M.; Reis, R.L.; Duarte, A.R.C. Natural Deep Eutectic Solvents—Solvents for the 21st Century. *ACS Sustain. Chem. Eng.* **2014**, *2*, 1063–1071. [CrossRef]

84. Wahlström, R.; Rommi, K.; Willberg-Keyriläinen, P.; Ercili-Cura, D.; Holopainen-Mantila, U.; Hiltunen, J.; Mäkinen, O.; Nygren, H.; Mikkelson, A.; Kuutti, L. High Yield Protein Extraction from Brewer's Spent Grain with Novel Carboxylate Salt-Urea Aqueous Deep Eutectic Solvents. *ChemistrySelect* **2017**, *2*, 9355–9363. [CrossRef]

85. Huang, J.; Guo, X.; Xu, T.; Fan, L.; Zhou, X.; Wu, S. Ionic deep eutectic solvents for the extraction and separation of natural products. *J. Chromatogr. A* **2019**, *1598*, 1–19. [CrossRef]

86. Bleakley, S.; Hayes, M. Algal Proteins: Extraction, Application, and Challenges Concerning Production. *Foods* **2017**, *6*, 33. [CrossRef]

87. Nadar, S.S.; Rao, P.; Rathod, V.K. Enzyme assisted extraction of biomolecules as an approach to novel extraction technology: A review. *Food Res. Int.* **2018**, *108*, 309–330. [CrossRef]

88. Boussetta, N.; Vorobiev, E. Extraction of valuable biocompounds assisted by high voltage electrical discharges: A review. *Comptes Rendus Chimie* **2014**, *17*, 197–203. [CrossRef]

89. Boussetta, N.; Vorobiev, E.; Reess, T.; De Ferron, A.; Pecastaing, L.; Ruscassié, R.; Lanoisellé, J.-L. Scale-up of high voltage electrical discharges for polyphenols extraction from grape pomace: Effect of the dynamic shock waves. *Innov. Food Sci. Emerg. Technol.* **2012**, *16*, 129–136. [CrossRef]

90. González-García, E.; Marina, M.L.; García, M.C. Nanomaterials in Protein Sample Preparation. *Sep. Purif. Rev.* **2019**, *49*, 1–36. [CrossRef]

91. Magalhães, L.M.; Segundo, M.A.; Reis, S.; Lima, J.L.F.C. Methodological aspects about in vitro evaluation of antioxidant properties. *Anal. Chim. Acta* **2008**, *613*, 1–19. [CrossRef] [PubMed]

92. Floegel, A.; Kim, D.-O.; Chung, S.-J.; Koo, S.I.; Chun, O.K. Comparison of ABTS/DPPH assays to measure antioxidant capacity in popular antioxidant-rich US foods. *J. Food Compos. Anal.* **2011**, *24*, 1043–1048. [CrossRef]

93. Vásquez-Villanueva, R.; Muñoz-Moreno, L.; José Carmena, M.; Luisa Marina, M.; Concepción García, M. In vitro antitumor and hypotensive activity of peptides from olive seeds. *J. Funct. Foods* **2018**, *42*, 177–184. [CrossRef]

94. Puchalska, C.; Esteve, M.L.; Marina, M.C. García Peptides. In *Handbook of Food Analysis*; CRC Press: Boca Raton, FL, USA, 2015; pp. 329–355, ISBN 978-1-4665-5654-6.

Review

Whey for Sarcopenia; Can Whey Peptides, Hydrolysates or Proteins Play a Beneficial Role?

Sarah Gilmartin [1,2], **Nora O'Brien** [2] **and Linda Giblin** [1,*]

1   Teagasc Food Research Centre, Moorepark, Fermoy, P61 C996 Co. Cork, Ireland; Sarah.Gilmartin@teagasc.ie
2   School of Food and Nutritional Science, University College Cork, T12 YN60 Co. Cork, Ireland; nob@ucc.ie
*   Correspondence: Linda.Giblin@teagasc.ie

Received: 6 May 2020; Accepted: 2 June 2020; Published: 5 June 2020

Abstract: As the human body ages, skeletal muscle loses its mass and strength. It is estimated that in 10% of individuals over the age of 60, this muscle frailty has progressed to sarcopenia. Biomarkers of sarcopenia include increases in inflammatory markers and oxidative stress markers and decreases in muscle anabolic markers. Whey is a high-quality, easily digested dairy protein which is widely used in the sports industry. This review explores the evidence that whey protein, hydrolysates or peptides may have beneficial effects on sarcopenic biomarkers in myoblast cell lines, in aged rodents and in human dietary intervention trials with the older consumer. A daily dietary supplementation of 35 g of whey is likely to improve sarcopenic biomarkers in frail or sarcopenia individuals. Whey supplementation, consumed by an older, healthy adult certainly improves muscle mTOR signaling, but exercise appears to have the greatest benefit to older muscle. In vitro cellular assays are central for bioactive and bioavailable peptide identification and to determine their mechanism of action on ageing muscle.

Keywords: sarcopenia; whey protein; muscle; C2C12; aged animals; older adult; exercise

1. Introduction

From 1980 to 2019, the number of people over the age of 65 worldwide doubled to 810 million people. This populace will reach 2 billion people by 2050 [1]. With a rapidly aging population, there is a need to understand how dietary intervention can counteract the physical impediments of the ageing process on muscle. The reduction in the sum of muscle fibers and size, in parallel with the deficit in spinal motor neurons, results in weakened mechanical muscle ability [2]. This, in turn, affects balance, gait and overall ability to perform tasks of daily living such as rising from a chair unassisted or the ability to walk independently [3]. Our percentage of muscle mass declines after the age of 30 at a rate of 3%–5% every 10 years and this decline accelerates after the age of 60 [4]. A decline in muscle mass and strength may eventually result in an individual presenting with a muscle mass lower than two standard deviations of the adult population mean and having a gait speed of <0.8 m/s [5]. At this juncture, a clinical diagnosis of sarcopenia is made [6]. By 2045, the incidence of sarcopenia in Europe will increase from 19 m (2016 figures) to 32 m—a 68% increase [7]. Shafiee et al. [8] estimated the incidence of sarcopenia in adults over the age of 60 at 10% based on the assessment of 58,404 individuals using the European Working Group on Sarcopenia in Older People (EWGSOP) [9], the International Working Group on Sarcopenia (IWGS) [10] and the Asian Working Group for Sarcopenia (AWGS) [11] definitions. A dual-energy X-ray absorptiometry scan (DEXA) is the preferred method used in the diagnosis of sarcopenia [12]. It evaluates fat mass together with bone mass, albeit its inability to decipher between water retention and fat infiltration in muscle can result in an 8% overestimation of skeletal muscle [13]. Other methods used to measure muscle mass include, bioelectrical impedance, neutron activation assessments and urinary excretion of creatinine [6]. In muscle cells, creatine to phosphocreatine acts as an important phospho energy store and is mediated by creatine kinase [14]. Since 90% of the body's phosphocreatine is stored in muscle tissue [15], circulating levels of its breakdown product, creatinine, is regarded as a reliable biomarker for muscle mass (where kidney function is normal) [16]. For 82 sarcopenic individuals, Rong et al. [17] observed that serum creatinine levels (66.68 ±14.21 μmol/L vs. 73.16 ±11.73 μmol/L, $p < 0.05$) were significantly lower than 82 non-sarcopenic individuals of similar age. Although further analysis with the dataset using univariate regression only indicated a tendency to associate with sarcopenia ($p = 0.058$), other studies have also observed significant associations between the serum creatinine pathway and sarcopenia [18,19].

1.1. Sarcopenia Associated Blood Biomarkers-Inflammatory Cytokines, Hormones, Muscle Anabolic Signals and Oxidative Stress Indicators

Although blood biomarkers are not used to diagnose sarcopenia, recent studies have shown significant differences in a range of other blood biomarkers between control and sarcopenic individuals of similar ages. Not surprisingly, circulating levels of several inflammatory cytokines have been associated with sarcopenia, indicating an inflammasome role in sarcopenia [20]. Most notably, interleukin-6 (IL-6) was significantly higher in those diagnosed with sarcopenia [17,18,21] with Rong et al. [17] reporting serum levels of 43.80 ± 10.13 pg/mL ($n = 82$) compared to 27.38 ± 9.53 pg/mL in the control group ($p < 0.05$) and Bian et al. [18] reporting 49.77 ± 22.14 pg/mL serum IL-6 in 79 individuals vs. 39.72 ± 29.53 pg/mL in the control group ($p = 0.03$). Univariate logistic regression analysis highlighted increased IL-6 as a risk factor for sarcopenia [21], with an increase in plasma IL-6 associating with a slower walking speed in the older adult ($n = 854$ mean age of 74.3 ± 2.7 years) [22]. However, it is controversial, with a meta-analysis performed on 17 studies with 3072 sarcopenic individuals reporting that serum IL-6 levels were similar in sarcopenic and non-sarcopenic participants [23]. Interleukin 10 (IL-10) is also regarded as a biomarker of note with elevated levels in serum of sarcopenic individuals (4.13 ± 1.03 pg/mL compared to 3.75 ± 1.21 pg/mL) compared to the control group [17]. However, Kwak et al. [21] observed no differences in plasma IL-10 levels in 50 sarcopenic vs. 46 non-sarcopenic individuals and Stowe et al. [24] reported no increase with age in 1411 individuals. In the study by Bian et al. [18], Tumor Necrosis Factor-α (TNF-α) was significantly elevated in serum collected from sarcopenic individuals (165.39 ± 19.49 pg/mL) compared to controls (148.79 ± 26.06 pg/mL, ($p = 0.01$)). In vastus lateralis biopsies, TNF-α mRNA transcript was 2.8 fold higher in older men ($n = 16$, 70 years old) compared to 13 men aged 20 [25], albeit the older individuals were not diagnosed with sarcopenia. Plasma concentrations of TNF-α do indeed increase as we age [24] but Kwak et al. [21] did not observe differences in sarcopenic vs. non-sarcopenic individuals over the age of 60 and a meta-analysis study by Bano et al. [23] study did not identify TNF-α as a biomarker of interest. The acute phase protein C-reactive protein (CRP), however, has received some considerable attention. Bano et al. [23] stated that 3072 sarcopenic individuals had significantly higher levels of circulating CRP (SMD = 0.51; 95%CI 0.26, 0.77; ($p < 0.0001$); I2 = 96%) than 8177 controls. Taaffe et al. [22] associated increased levels of this hepatic inflammatory protein in serum with reduced grip strength. The pro-inflammatory cytokine macrophage migration inhibitory factor (MIF) was also identified by Kwak et al. [21] as one of four biomarkers from a total of 21 blood biomarkers that could be used in a screening panel for sarcopenia with significantly higher levels in the sarcopenic participants (25.1 ± 1.19 vs. 20.71 ± 0.89 ng/mL, ($p = 0.008$)). The four biomarkers were IL-6, MIF, an extracellular matrix repair glycoprotein secreted protein acidic and rich in cysteine (SPARC), and insulin-like growth factor 1 (IGF-1). IGF-1 involvement in sarcopenia may not be a surprise as IGF-1 plays an important role in muscle protein synthesis and is known to decrease with age [26]. IGF-1 participates in muscle anabolism via Akt phosphorylation activating the mammalian target of rapamycin (mTOR) pathway which ultimately controls muscle protein synthesis and turnover [27]. Sarcopenia results from a decrease in muscle anabolic pathways with an increase in catabolic pathways [28]. For sarcopenic individuals, Kwak et al. [21] observed a significant (further) decrease in serum IGF-1 from 72.61 ± 5.49 ng/mL to 58.16 ± 3.37 ng/mL compared to the non-sarcopenic control group of similar age. Oxidative stress due to free radical damage has been suggested as one of the most prominent causes of skeletal muscle reduction that occurs with ageing [29]. These radicals are exceedingly reactive species with the ability, either in the nucleus or in cell membranes, of damaging DNA, proteins, carbohydrates and lipids [30]. To mop up free radicals and maintain redox homeostasis, the muscle cell employs enzymes such as glutathione peroxidase (GPx) [31], superoxide dismutase (SOD) [32] and catalase (CAT) [33]. Studies have shown that oxidative damage biomarkers, 8-hydroxy-2′deoxyguanosine (8-OHdG) and malondialdehyde (MDA) are increased in skeletal muscle as we age [34].

1.2. Why Whey?

Skeletal muscle (SM) makes up approximately 40% of total body weight, consisting of 50%–75% of total proteins, and is accountable for 30%–50% of total body protein turnover [35]. Dietary intervention with a high-quality protein could modulate the biomarkers of ageing muscle and/or delay the onset of sarcopenia [36]. Bovine whey is used, in a wide variety of products (beverages and protein bars), in the sports nutrition market to promote muscle growth [37] and repair [38] after physical exercise [39]. Its success is predicted to be due to the fact that whey proteins (1) are easily digestible [40,41], (2) contain all essential amino acids [42], (3) are a rich source of branch-chain amino acids (BCAA) [43] which activates the mTOR pathway [44,45], and (4) are a rich source of bioactive peptides [46]. Bovine whey is composed of β-lactoglobulin (50%–60%), α-lactalbumin (15–25%), bovine serum albumin (BSA, 6%), lactoferrin (<3%) and immunoglobulins (<10%) [47]. Bovine whey is used in food formulation in different formats that differ in their protein concentration and degree of protein hydrolysis; i.e., liquid whey, whey protein isolate, whey protein concentrate or whey protein hydrolysate [48].

This review considers the evidence that whey peptides, hydrolysates, proteins or products can delay or reduce symptoms of sarcopenia or alter biomarkers of sarcopenia in the older adult, in aged animals or muscle cells lines.

2. Whey Peptides on Muscle Cell Lines In Vitro

Treatment of murine myoblast cell line, C2C12, with whey peptides, hydrolysates or intact protein is summarised in Table 1. Previously, our group [49] has identified several whey peptides that are produced during simulated upper gastrointestinal digestion of whey protein isolate. A subgroup of these peptides was capable of crossing the intestinal barrier, as evidenced by their appearance on the basolateral side of Caco2-HT29 monolayers. Within this subset, four peptides ALPM, GDLE, VGIN and AVEGPK (5 mM) reduced oxidative stress in undifferentiated C2C12 cells [49]. Whether these peptides can be detected in plasma after dietary intervention with whey is as yet unknown. Ogiwara et al. [50] identified a dipeptide (MH), wheylin-1, from β-lactoglobulin produced during thermolysin enzymatic hydrolysis. This peptide significantly increased insulin-induced Akt phosphorylation in differentiated C2C12 cells compared to insulin induction alone or compared to control cells. Whether this peptide can cross the intestinal barrier is unknown. Certainly, intraperitoneal injection of wheylin-1 (1 mg/kg body weight) in young mice ($n - 5$) increased Akt phosphorylation in skeletal gastrocnemius muscle. Its effect in dietary intervention trials on aged animals or older/sarcopenic humans has not been tested. Mobley et al. [51] differentiated C2C12 cells prior to treatment with whey hydrolysate and surprisingly noted a significant decrease in mRNA transcripts of the mTOR biomarker, raptor (a regulatory associated protein [52]), compared to levels in cells in DMEM media alone. The peptide or amino acid composition of the whey hydrolysate was not given. Kerasioti et al. [53] pre-incubated differentiated C2C12 cells with sheep whey protein for 24 h and then oxidatively stressed the cells with tertbutyl hyrdoperoxide for 30 min. Glutathione (GSH) levels significantly increased and ROS decreased with treatment of sheep whey protein at 1.56, 3.12 and 6.24 mg/mL, albeit the description (intact/hydrolysate) of the whey protein was not provided. Xu et al. [54] also measured levels of GSH, SOD, CAT and G-Px to evaluate the protective effects of whey protein on undifferentiated C2C12 cells from oxidative damage. Cells pre-treated with whey protein for 24 h and then subjected to hydrogen peroxide stress, resulted in significant decreases of MDA levels and significant increases in GSH,

SOD, CAT and G-Px levels compared to hydrogen peroxide stressed cells. Knight et al. [55] purified the minor protein, Ribonuclease 5, from bovine whey and demonstrated that this purified fraction (95% pure) not only significantly increased myogensis of C2C12 cells over a 4-day period but also increased C2C12 creatine kinase activity. Most notably, these increases were similar to IGF-1 treatment. However, it is important to question the rational for applying intact whey proteins directly to muscle cells as only a limited number of whey peptides will survive the proteolytic conditions of the gut and cross the intestinal barrier to reach muscle. Knight et al. [55] did report a dietary intervention trial in adult, but not aged mice, where grip strength and muscle weight were significantly increased after 3 weeks and 4 months, respectively, with a diet supplemented with Ribonucelase 5 (50% purity, 17 ug/g feed). Indeed, Carson et al. [56] collected serum from six healthy men 60 min after ingesting whey protein hydrolysate (0.33 g/kg body weight). C2C12 myotubes were then treated with this serum in the presence of media for 4 h. Phosphorylation to total protein ratios were increased for mTOR, p70S6K, 4EBP1 in media conditioned with serum post whey consumption vs. serum from fasted state. This increased phosphorylation status was not observed when cells were incubated with serum collected from individuals who consumed an equivalent non-essential amino-acid-based beverage.

Table 1. Effects of whey protein supplementation on muscle cells in vitro. BSA: Bovine serum albumin; ABAP: 2,2′-azobis(2-methylpropionamidine) dihydrochloride; Akt: protein kinase B; DMEM: Dulbecco's Modified Eagle Medium; tBHP: tert-butyl hydroperoxide; GSH: glutathione; ROS: reactive oxygen species; H_2O_2: hydrogen peroxide; MDA: malondialdehyde; SOD: superoxide dismutase; CAT: catalase; G-Px: glutathione peroxidase; IGF-1: insulin-like growth factor 1.

Whey Protein Description	Experimental Detail	Outcome	Reference
ALPM (β-lactoglobulin 142-145AA) GDLE (β-lactoglobulin 52-55AA) VGIN (α-lactoalbumin 99-102AA) AVEGPK (BSA 568-573AA)	Undifferentiated C2C12,-5 mM synthesized peptide pretreated for 1 h, 37 °C prior to ABAP radical (600 μM)	Each peptide ↓ % free radical formation similar to control V cells treated with ABAP treatment	[49]
Wheylin-1 (MH), thermolysin hydrolysis β-lactoglobulin	Differentiated C2C12, 500 μM synthesized peptide for 3 h at 37 °C, plus insulin (100 nM) for 15 min	↑ phosphorylated Akt: Akt total protein V cells treated with insulin alone or V saline	[50]
Hydrolysed whey protein (COMBAT-MusclePharm, proprietary blend from whey protein concentrate)	Differentiated C2C12, 6 h treatment with 13 μg/mL	↓ raptor mRNA V cells treated with DMEM media alone	[51]
Sheep whey protein	Preincubated differentiated C2C12 cells with sheep whey protein for 24 h at 0.78 1.56, 3.12 and 6.24 mg/mL. Stressed with tBHP for 30 min	↑ GSH, ↓ ROS with 1.56, 3.12 and 6.24 mg/mL V cells treated with tBHP alone	[53]
Whey protein	Undifferentiated C2C12 whey protein (0.4 mg/mL) for 24 h and then stressed with 0.5 mM H_2O_2	↓ MDA V cells treated with H_2O_2 control ↑ GSH, SOD, CAT, G-Px V cells treated with H_2O_2 control	[54]
Ribonuclease5 enriched whey from bovine skim milk	Differentiated C2C12,10 μg/mL Ribonuclease 5 fractions that differed in purity, 4-day differentiation	↑ myogenesis myosin heavy chain staining V cells treated with media alone ↑ creatine kinase activity similar to IGF-1 (200 ng/mL) positive control at 3 and 4 days V cells with media alone	[55]

3. Effects of Whey Products on Aged Animal Models

Rats and mice over the aged of 24 months are considered equivalent to humans of >60 years of age [57,58]. Recently, several intervention trials with whey have been performed on aged rodents, (Table 2). Many of these studies included whey dietary intervention in combination with BCAA, leucine or with potent antioxidants. Van Dijk et al. [59] fed 22-month-old mice for 3 months with a whey-based diet or a whey-based diet supplemented with antioxidants selenium, zinc, vitamin A and vitamin E. Dietary intervention with whey or antioxidants had no effect on lean mass compared to control group but in combination they significantly increased lean body mass. Maximal in vivo muscle strength was significantly increased in mice fed whey or whey plus antioxidants compared to control diets. Nocturnal physical activity was higher in the whey treatment group. Mosoni et al. [60] compared two different doses of whey-based diets, whey protein (12%) or (18%) with/without anti-inflammatory/antioxidant mix using 16-month-old rats over a period of 6 months. Lean body mass % loss associated with ageing was less with 18% whey supplementation than with 12% whey. There was no difference in vivo muscle fractional synthesis rate across whey dosage or casein control diets. Ex vivo protein synthesis rate and proteolysis rate were measured in biopsied epitrochlearis muscle. Ex vivo synthesis rate with whey was significantly increased over casein diet and there was no difference with ex vivo proteolysis rate. Supplementation with anti-inflammatory/antioxidant mix in both diets significantly reduced the redox biomarker, thiobarbituric acid reactive substances in muscle, while increasing muscle GSH and plasma antioxidant activity. In contrast, casein diet-improved liver SOD levels compared to whey. Plasma fibrinogen, an inflammatory biomarker was elevated after 6 months in all groups compared to time zero. The authors viewed this increase as an indicator of ageing in the animals. The higher dose of whey provided a protective effect from this ageing-associated inflammation, with fibrinogen significantly less in the group fed 18% whey compared to those fed 12% but not casein. Garg et al. [61] investigated the ability of whey protein concentrate to counteract the effects of oxidative stress in 24-month-old aged male rats. Rats were fed whey protein concentrate orally for 28 days. Erythrocytes were harvested and structural and functional integrity of their cell membranes examined. Membrane integrity deteriorates with age with notable decreases in sialic groups and increases in carbonyls and lipid peroxidation. Garg et al. [61] observed that the membrane sulfhydryl groups and sialic acid groups were significantly increased, whereas lipid hydroperoxide and protein carbonyls were both significantly decreased in erythrocytes membranes of old rats fed whey protein concentrate in comparison to aged controls.

Table 2. Aged animal intervention trials outlining the effects of whey supplementation. SOD: superoxidase dismutase; GSH: glutathione;FSR: fractional synthesis rate; Akt: protein kinase B; FFM: fat free mass.

Whey Dosage, Source, Duration	Animal Model	Outcome of Whey Intervention	Reference
Whey (136 g/kg feed) + leucine (16.8 g/kg feed) 3-month intervention Isocaloric diets	22-month-old maleC57/BL6J mice $n = 8$	Whey: No difference lean body mass, ↑ fore limb strength V diet low in antioxidants, ↑ nocturnal physical Activity V control diet ↓ fatigue V control diet, tendency ($p < 0.07$) for improved muscle quality (grip strength/lean mass) V diet low in antioxidants	[59]
Whey protein (140 g/kg feed)(12%) or (215 g/kg feed)(18%) with/without anti-inflammatory /antioxidants mix (chamomile extract 17 g/kg feed, Vitamin E 300 UI/Kg feed and Vitamin D 5000 UI/Kg feed) 6 months	16-month-old male Wistar rats $n = 11$	Whey: No difference in vivo gastrocnemius muscle fractional synthesis rate synthesis or degradation rates Lean body mass loss through ageing ↓ with whey over time 12% and 18% ↑ ex-vivo epitrochlearis muscle synthesis rate V casein ↓ SOD in whey V casein diet ↑ GSH liver in whey + antioxidants V whey - antioxidants ↓ plasma fibrinogen whey 18% V whey 12% but not to casein ↑ plasma antioxidant activity whey V casein diets No difference in vivo muscle FSR across whey dosage or casein control diets Ex vivo synthesis rate with whey was ↑ V casein diet	[60]
Whey protein concentrate (Camillotek), 0.300 g/kg body weight 28 days by oral gavage daily	24-month-old male Wistar rats $n = 6$	Whey:↑ erthyrocyte membrane sulfhydryl groups and sialic acid groups V saline oral gavage ↓ lipid hydroperoxide and protein carbonyls in erthyrocyte membrane V saline oral gavage	[61]
Oral gavage (0.5 mL of 0.139 g whey included 0.188 g leucine) whey protein isolate plus leucine (Lacprodan9224 Arla Foods) Post-prandial study 60, 75, 90 min	25-month-old male C57/BL6J mice $n = 8$	Whey:↑ muscle protein synthesis V fasted or leucine alone ↑ phosphorylated Akt and phosphorylated 4E-BP1 V fasted or leucine alone ↑ phosphorylated p70S6kV fasted	[62]

Table 2. *Cont.*

Whey Dosage, Source, Duration	Animal Model	Outcome of Whey Intervention	Reference
6 g meal (0.864 g whey) Post prandial study 0, 90, 125, 180, 240 min post ingestion	20-month-old male Wistar rats $n = 10$ per time point	Whey: ↑ muscle protein synthesis (MPS) rate V casein group (diet X time interaction for whey V soy group) ↑ plasma leucine V casein and soy group ↑ phosphorylated Akt, S6K1 and S6: total Akt, S6K1 and S6 S ratio V casein and soy groups with some diet X time interactions	[63]
Whey protein (160 g/kg feed) (Lactalis) 5-month intervention ad libitum and energy-restricted diets	21-month-old male Wistar rats $n = 10$	Ad libitum + whey: Tendency ↑ soleus muscle weight V casein diet Ad libitum + whey and restricted + whey: ↑ muscle protein absolute synthesis rate V casein diet Energy restricted + whey: tendency ↑ muscle strength V casein diet	[64]
Whey protein 0.85 g bolus, 5 days/week over 2 months (+/− exercise)	17-month-old male Wistar rats $n = 16$	Whey: ↑ hind limb stride length of sedentary rats V time zero ↑ maximum voluntary walking speed of active rats V time zero No difference in total movement, distance travelled, activity time or average speed for whey V casein or milk protein intervention ↓ stance time of whey active rats V time zero ↓ brake time of whey active rats V time zero No difference in absolute or relative grip force	[65]
Unilateral hindlimb casting for 8 days followed by 40-day recovery with 144 g/kg feed whey (Prolacta, Lactalis)	22-old-male Wistar rats $n = 17$	Whey: Day20, Day 40 ↑ muscle mass gain V day-1 recovery, control diet and leucine-rich diet Day 40 ↑ muscle protein synthesis V day-1 recovery, control diet and leucine-rich diet	[66]

Dijk et al. [62] favored the combination of whey protein isolate with leucine in a post-prandial study with 25-month-old mice. The rate of muscle protein synthesis was measured in response to oral gavage with whey protein isolate plus leucine compared to oral gavage with leucine or a fasted control. Interestingly, the whey and leucine combination resulted in a significant increase in muscle protein synthesis after 60 min post oral gavage in comparison to the control groups. Phosphorylation of Akt and other mTOR signalling proteins (4E-BP1 and p70S6k) were significantly elevated at 60 min post-ingestion of whey and leucine compared to leucine alone (Akt, 4E-BP1) or the fasted state (Akt, 4E-BP1 and p70S6k). Jarzaguet et al. [63] fed aged rats (20 months) whey proteins 144 g/kg feed containing leucine 16.2 g/kg. Phosphorylation status of Akt and other mTOR biomarkers (S6K1 and S6) were significantly elevated after ingestion of whey bolus compared to casein or soy with some diet by time interactions. Muscle protein synthesis rate was also significantly increased compared to casein intervention with time by diet interactions noted with soy diet. Walrand et al. [64] examined the effects of whey protein on muscle weight, strength and protein synthesis in ad libitum diets, protein-restricted or energy-restricted diets in 21-month-old aged rats for 5 months. Whey protein enhanced muscle absolute synthesis rate for all three diets. However, only when diets were restricted did whey intervention have a tendency to have a positive effect on skeletal muscle strength. There was no significant difference noted between the weights of the soleus muscle across the groups but the muscle weight tended to be higher in the ad libitum whey protein group compared to the ad libitum casein group (266.5 mg vs. 238.7 mg, respectively) [64]. Interestingly, Lafoux et al. [65] favored a combination of whey with exercise in their study on aged rats. Seventeen-month-old rats were fed a bolus of 0.85 g of whey or casein or milk protein and were subjected to either a sedentary or active routine over a 2-month period. Although there were some notable differences compared to time zero (Table 2), total movement, distance travelled activity time or average speed did not differ between whey or treatment groups. Interestingly, Magne et al. [66] focused on muscle recovery in the older animal. The experimental design included immobilisation of the hind limbs of aged rats (22–24 months) for 8 days. Upon casting removal, the animals were fed for 40 days on a diet which contained whey protein at 144 g/kg feed. Muscle mass gain in animals that received whey was significantly increased by ~200 mg by day 20 and further increased to ~400 mg by day 40 compared to casein- or leucine-rich diet groups. An increase in postprandial muscle protein synthesis and amino acid concentration were both noted after 40 days with whey compared to a casein diet, to day 1 of recovery or to a leucine-rich diet.

All of the rodent studies observed a beneficial effect of whey supplementation on one or more biomarkers of interest (e.g., antioxidant, muscle protein synthesis and muscle Akt phosphorylation levels). In terms of dosage, it is interesting to note that in murine studies, mice received 136 g whey per kg feed, which equates to 0.54 g whey per day, assuming an older mouse consumes approximately 4 g feed. In studies with aged rats, whey dosage was 140–215 g whey per kg feed, which equates to 2.9–4.5 g whey/day if an older rat consumes 21 g feed per day [60]. Bolus and oral gavage experiments were performed with 0.85, 0.864 or 0.15 g whey for rats and 0.139 g whey for mice. In addition, animal trials were performed with whey protein isolate, whey protein concentrate or undefined whey with no mention of whey hydrolysates or individual whey peptides.

4. Intervention Trials with Whey and the Older Adult

In humans, a number of postprandial studies (Table 3) with whey supplementation have been performed in the older adult. The direct method to track the fate of whey post-ingestion is to label it. Pennings et al. [67] investigated the effects of radiolabelled whey protein in 33 healthy, elderly men aged 73 ± 2 yrs. An infusion of L-[1-^{13}C] phenylalanine to a lactating Holstein cow resulted in radiolabelled milk from which the whey protein fraction was purified. The 73-year-old man received 10, 20 or 35 g of this radiolabelled whey. At 240 min, muscle biopsies were collected from the vastus lateralis and muscle tissue analysis was performed. Whole body protein breakdown was significantly decreased in the groups post whey protein consumption when compared to time zero. Synthesis, oxidation and net balance of proteins were all significantly increased compared to time zero after ingestion of 35 g of whey protein. Overall, net balance of protein metabolism was significantly higher for the 35 g whey protein supplemented group compared to the 20 g whey protein group. Muscle biopsies also revealed dose-dependent increases in muscle fractional synthesis rate and incorporation of radiolabelled phenylalanine. Reitelseder et al. [68] performed a postprandial study combining a 300 mL whey protein hydrolysate beverage (0.45 g/kg) with exercise over a 6 h period in older men (aged 61 ± 1 year) (n = 10). A continuous infusion by arm vein catheterization of radioactive free L [^{15}n]-phenylalanine tracer allowed for protein-bound phenylalanine measurement in biopsied vastus lateralis muscle. Whey increased muscle protein synthesis compared to basal control but was similar to casein or carbohydrate intervention. In addition, mTOR phosphorylation status was unaffected in this acute study. With limited effects reported, the authors questioned whether or not older muscle would respond to an increase in dietary intake of whey protein. Borack et al. [69] combined whey protein isolate with exercise and examined the effects of 30.4 g of whey protein isolate in elderly individuals (55–75 years of age) on phosphorylation of mTOR biomarkers and fractional protein rates in vastus lateralis biopsies. Although consumption of whey increased muscle mTOR phosphorylation, increased fractional protein synthesis and decreased breakdown compared to baseline, there was no difference with consumption of a soy-dairy protein blend. Wilkinson et al. [70] compared the effects of whey protein (40 g bolus) with or without exercise on muscle protein synthesis in older women aged 65 ± 1 year (n = 24) in a 7 h postprandial study. Muscle myofibrillar protein fractional synthesis rate (FSR) significantly increased after 2 h with whey supplementation plus exercise compared to time zero but was similar to the 6 g of leucine plus exercise group. Interestingly, phosphorylation of mTOR protein, p-p70S6K1 was significantly increased with whey compared to time zero. However, exercise was required to increase phosphorylation of p70S6K1 in both the leucine and whey treatment groups vs. time zero. Kramer et al. [71] performed a postprandial study to investigate the effects of leucine-enriched whey protein supplementation (3 g of leucine/21 g of whey) in healthy (69 ± 1 years old) (n = 15) and sarcopenic (81 ± 1 years old) (n = 15) males. This study also included a continuous infusion by arm vein catheterization of radioactive free L-[ring-^{15}C$_6$]-phenylalanine tracer to allow for protein-bound phenylalanine measurement in muscle (vastus lateralis). Muscle protein-bound enrichment gain was higher in the sarcopenic group at time zero after ingestion of whey protein and increased significantly in both sarcopenic and healthy control groups over time. Mixed muscle protein synthesis rate was significantly increased postprandially compared to basal values following whey protein supplementation, regardless of whether an individual was sarcopenic or healthy. Dideriksen et al. [72] examined the effects of 0.45 g/kg LBM whey protein isolate plus resistance training in 14 older men aged ≥ 60 years of age with a high level of plasma CRP (>2 mg/L). Seven subjects were given 1800 mg/day of ibuprofen in tandem with whey and [^{15}N] phenylalanine stable isotope tracer. A further seven subjects received whey protein and a placebo tablet containing potato starch and lactose monohydrate. Myofibrillar FSR was elevated following intake of whey protein in both resting and exercise legs in comparison to time zero. Myofibrillar FSR was higher in the exercised leg compared to the resting leg. There was a significant difference for all three groups for basal, postprandial and post-exercise states but there was no difference between the control (untreated), whey or whey plus ibuprofen groups themselves. For those in the whey plus exercise groups, muscle connective tissue was increased compared to time zero. Smith et al. [73] measured phosphorylation status of mTOR^{Ser2448}, AKT^{Ser473}, and AKT^{Thr308} proteins in 22 women (aged 57.8 ± 4.2 years) during a hyperinsulinemic-euglycemic clamp procedure with/without whey protein. Phosphorylation of mTOR^{Ser2448} and p70S6K^{Thr389} were greater during whey protein supplementation compared to Kool Aid controls and AKT^{Ser473} and AKT^{Thr308} were increased with whey compared to time zero.

Table 3. Postprandial trials in the older adult with whey intervention. mTOR: mammalian target of rapamycin; FSR: fractional synthesis rate; LEAA: leucine essential amino acid; WP: whey protein; * power calculation to determine number of subjects was performed.

Cohort	Whey Source, Dosage, Duration	Outcome of Whey Supplementation	Reference
$n = 33$ aged 73 ± 2 years, males Randomised	Postprandial L-[1-^{13}C] phenylalanine-labeled whey protein 10 g, 20 g, or 35 g containing Givaudan Whey $n = 11$	↓ Whole body protein breakdown V time zero 35 g or 20 g ↑ whole protein synthesis V 10 g whey 35 g ↑ whole protein oxidation V time zero and 10 g whey Dose dependent ↑ whole protein net balance 35 g ↑ muscle fractional protein synthesis rate V 10 g or time zero Dose-dependent ↑ radiolabeled phenylalanine muscle incorporation	[67]
$n = 27$, aged 61 ± 1 years, males Randomised, isocaloric controlled	Postprandial 300 mL whey protein hydrolysate beverage (0.45 g/kg lean body mass) (PEPTIGEN Arla foods) Exercise = leg extensions Whey $n = 10$	↑ Myofilbrillar muscle protein synthesis in rested muscle biopsies V fasted state but similar to casein or maltodextrin control No difference phosphorylated mTOR p70S6K: V casein or maltodextrin in rested or exercised state	[68]
$n = 20$ aged 55–75 years, males Randomised, double blinded, controlled	Postprandial Whey protein isolate 30.4 g (DuPont Nutrition and Health) Exercise = leg extension Whey $n = 10$	↑ Muscle fractional synthesis rates FSR V baseline but no difference to soy-dairy protein blend ↓ Muscle fractional breakdown rate V baseline but no difference to soy-dairy protein blend ↑ mTORC1, S6KI, 4E-BP1, rpS6 V baseline no difference to soy-dairy protein blend	[69]
$n = 24$, aged 65 ± 1 year, females Randomised	Postprandial Whey protein 40 g (Ajinomoto Co.) Exercise = knee extensions Whey $n = 8$	↑ Muscle myofibrillar protein fractional synthesis rate after 2 h for 1.5 g, 6 g LEAA WP and after 4 h for 6 g LEAA + WP V time zero ↑ Muscle myofibrillar protein fractional synthesis rate after 4 h for all t1.5 g, 6 g LEAA and WP + exercise V time zero ↑ Phosphorylated p70S6K1 V time zero ↑ Phosphorylated p70S6K1 for 6 g LEAA and WP + exercise after 2 h and only WP + exercise after 4 h V time zero	[70]

Table 3. *Cont.*

Cohort	Whey Source, Dosage, Duration	Outcome of Whey Supplementation	Reference
* $n = 30$ healthy (69 ± 1 years old) ($n = 15$) and sarcopenic (81 ± 1 years old) ($n = 15$) males Sarcopenia = gait speed ≤ 1.0 m/s, handgrip strength < 30 kg, SMMI < 8.4 kg/m^2	Postprandial Leucine enriched whey protein 3 g of leucine/21 g of whey (Nutricia advanced medical nutrition) Whey $n = 30$	↑ Muscle tissue free enrichment, muscle protein bound enrichment gain and mixed muscle protein synthesis rate in both healthy and sarcopenic individuals V time zero	[71]
$n = 24$, ≥60 years, males Randomised, cross-sectional, double blinded, placebo controlled.	Postprandial 0.45 g/kg LBM whey protein isolate containing 21.3–37.6 g protein (Lacprodan, Arla Foods) Exercise = knee extension Whey $n = 14$	↑ Myofibrillar FSR in resting and exercised legs V time zero ↑ Connective tissue FSR for whey + exercise V time zero Tendency towards a difference between postprandial & post exercise states for both whey and whey + ibuprofen groups	[72]
$n = 22$ aged 50 to 65 years postmenopausal women Randomised	Postprandial Whey protein trial consumed either 0.6 g whey protein per kg FFM (ProSynthesis Laboratories) Whey $n = 11$	↑ Phosphorylated mTOR^{Ser2448} & p70S6K^{Thr389} in muscle V time zero and control values (Kool Aid solution only) ↑ Phosphorylated AKT^{Ser473} and AKT^{Thr308} in muscle V time zeros	[73]

Exercise, specifically resistance or strength training has been shown to increase muscle mass in the elderly [74]. Whether whey in combination with resistance training (Table 4) can increase muscle mass and strength in a bid to combat sarcopenia has been investigated by a number of research groups. Nabuco et al. [75,76] examined the effects of 35 g of hydrolysed whey protein supplementation in older women aged > 60 years old over a 26-week period ($n = 47$). Whey protein consumption resulted in a decrease in uric acid compared to the maltodextrin group. Whey increased SOD, CAT and total radical-trapping antioxidant parameter (TRAP) vs. time zero. Whey resulted in a decrease in advanced oxidation protein products (AOPP) and lipid hydroperoxide compared to time zero. Whey supplementation resulted in a decrease in the 10 min walk speed test compared to time zero and maltodextrin placebo. A decrease in the time taken to rise from a seated position following whey consumption was also recorded compared to time zero. Whey supplementation resulted in an increase in knee extension, chest press and total strength compared to time zero and maltodextrin placebo. An increase in the arm exercise preachers curl was also noted following whey vs. time zero. Percentage of skeletal muscle mass was significantly increased following whey protein consumption compared to maltodextrin placebo. To try to unravel exercise from whey intervention, Sugihara Junior et al. [77] included a pre-conditioned 8-week resistance training followed by a 12-week whey intervention plus resistance training with women aged 67.4 ± 4 years old. The group supplemented with hydrolysed whey protein exhibited a significantly greater increase in chest presses, knee extensions, and total strength compared to maltodextrin placebo plus exercise and compared to 8-week exercise alone. Mori et al. [78] investigated the effects of 22.3 g of whey protein with and without exercise over a 24-week period in healthy elderly woman aged 65–80 years old ($n = 75$). Exercise (rising and sitting from a chair, plus leg extension with resistance band exercises) with whey was better than exercise alone which, in turn, was better than whey alone for muscle mass, grip strength, knee extension and gait speed. Interestingly, any intervention significantly improved these markers from time zero measurements. Englund et al. [79,80] examined the effects of whey protein supplementation (20 g) with Vitamin D (800 IU) and resistance training over a 6-month period in older adults aged 78.5 ± 5.4 years old ($n = 149$). Whey consumption with exercise resulted in an increase in normal density muscle compared to control (a low-calorie placebo drink 30 kcal, no protein, no Vitamin D) and exercise group. There was a significant decrease in low-density muscle in the whey protein group compared to baseline parameters and normal muscle density. Both whey plus exercise and control plus exercise groups showed a significant increase in knee flexor strength, power and quality compared to time zero. Whey protein supplementation and exercise resulted in a greater decline in intramuscular fat compared to the control plus exercise group. Chalé et al. [81] analyzed the effects of whey protein concentrate supplementation (40 g/daily) in 42 elderly adults (aged 70–85 years old) with resistance training for 6 months. Knee extensor power was the only significant increase following whey consumption compared to the maltodextrin control group. There was an observable increase in total mid-thigh cross-sectional area (CSA), total muscle CSA and total normal density muscle CSA in the whey protein group compared to time zero but no difference compared to maltodextrin. Kirk et al. [82] investigated the effects of leucine (0.03 g/kg/meal) enriched whey protein (0.5 g/kg/day) in elderly individuals (aged ≥60–86 years old) over a 16-week period in conjunction with exercise ($n = 46$). Significant increase in leg press, chest press and bicep curl were observed from pre- to post-intervention in both the whey protein in combination with exercise and exercise only groups. Ingestion of whey did not increase all fore-mentioned parameters over and above exercise alone, albeit whey plus exercise was better than time zero. In a follow-on study [83], rectus femoris and bicep femoris were less fatigued with exercise intervention but no additional improvement with whey compared to time zero. No differences were noted for muscle mass or handgrip between or within whey and exercise, exercise only, whey only or control (untreated) groups. Exercise alone appears to have a more profound effect on muscle health compared to whey and exercise or even whey alone. Hospitalized individuals are particularly vulnerable to muscle wastage. Whether whey protein supplementation can preserve muscle mass in this cohort was investigated by Niccoli et al. [84]. Frail patients aged 81.3 ± 1 years old (*n* = 47) were supplemented with 24 g of whey protein per day by incorporation in their porridge (9 g) and the remaining via dairy beverages (7.5 g whey/drink) at lunch and dinner. There was a significant improvement in grip strength (30.3%) and knee extensor force (42.7%) compared to time zero. Gait speed, and time to 'get up and go' for those supplemented with whey protein was significantly improved compared to pre-intervention parameters. Serum IL-6 was shown to significantly decrease following whey protein supplementation compared to time zero for both the whey-protein-supplemented group and the control group (hot cereal and milk products). Whey resulted in a greater decrease in percentage IL-6 compared to the control group (hot cereal and milk products without whey protein). Whether whey protein supplementation can aid in enhancing physical function following discharge from hospital was investigated by Gade et al. [85,86]. They examined the influence of whey protein supplementation and resistance training in older adults aged 70 years or older (*n* = 141) during their hospital stay and 12 weeks after discharge. Subjects were given a ready-to-drink milk-based protein supplement containing 27.5 g whey protein/day in combination with a resistance training programme over a 12-week period. Overall, there was no noted benefit (30 s chair stand test, hand grip strength, 4 m gait speed) of whey protein supplementation in this study. Mancuso et al. [87] investigated the effects of a whey-based oral supplement (10 g/day for 30 days) in 13 individuals suffering from mitochondrial disease aged 52.5 ± 15.2 years. Defective mitochondria can trigger a series of events that results in death of motor neuron and muscle fiber death, ultimately causing sarcopenia. Following treatment with whey advanced oxidation protein products (AOPP), a biomarker of oxidative stress in plasma was significantly decreased, while plasma FRAP and GSH were significantly increased during resting compared to the casein placebo group and baseline parameters.

Table 5 describes intervention trials in the older adult with whey supplementation where exercise was not included. Chanet et al. [88] reported on the effects of leucine-enriched whey protein (21 g) in combination with vitamin D (800 IU) in 12 healthy older men (71 ± 4 years old) over a 6-week period. Infusion by arm vein catheterization of radioactive free L-[^{2}H$_5$]-phenylalanine tracer was performed to allow for protein-bound phenylalanine measurement in muscle (vastus lateralis). Mixed muscle protein synthesis rate was elevated compared to the controls (flavored water) and to time zero. There was a significant increase in appendicular muscle mass and leg lean mass in the whey protein group compared to the flavored water group after 6 weeks. There was no difference noted between the whey protein group or the flavored water group for inflammatory biomarkers. Rodondi et al. [89,90] investigated the effects of whey protein supplementation (15 g) in combination with essential amino acids (5 g) with or without zinc (30 mg/day) in 47 elderly patients aged 85 ± 7.4 years for 4 weeks. With whey protein supplementation, activity of daily living score improved marginally suggesting a level of independent living by the subject. In the whey protein groups, IGF-1 levels were significantly increased vs. the controls (untreated). Coker et al. [91] examined the effects of a calorie-restricted whey-based meal replacer (170 kcal/ 5 time/day). Eleven obese elderly subjects aged 65–80 years consumed a whey protein (7 g) and essential amino acid (EAA) (6 g) meal replacement. There was an increase in skeletal muscle FSR in the whey group which was significantly different from the control group (competitive meal replacer).

Table 4. Whey and exercise intervention trials with the older adult. SOD: superoxide dismutase; CAT: catalase; TRAP: total radical-trapping antioxidant parameter; AOPP: advanced oxidation protein products; LST: lean soft tissue; CSA: cross-sectional area; IL-6: interleukin 6; WPS: whey protein supplementation; SPPB: Short Physical Performance Battery; FRAP: ferric reducing antioxidant power; GSH: glutathione; * power calculation to determine number of subjects was performed.

Cohort	Whey Source, Dosage, Duration	Outcome of Whey Supplementation	Reference
$n = 70 \geq 60$ years, females Randomised, double-blind, placebo-controlled	26 weeks 35 g hydrolyzed whey protein (Lacprodan, Arla Foods) Exercise = chest press, horizontal leg press, seated row, knee extension, preacher curl, leg curl, triceps pushdown, and seated calf raise Whey $n = 47$	↓ Uric acid V maltodextrin. ↑ SOD, CAT and TRAP V time zero ↓ AOPP and lipid hydroperoxide V time zero	[76]
$n = 70 \geq 60$ year, females Randomised, double-blind, placebo-controlled	26 weeks 35 g hydrolyzed whey protein (Lacprodan, Arla Foods) Exercise = chest press, knee extension & preacher curl Whey $n = 47$	↑ Total lean soft tissue, appendicular LST, lower LST & V time zero and maltodextrin ↓ 10-min walk speed and rising from seated position V time zero equal to maltodextrin ↑ Knee extension, preachers curls, chest press and total strength V time zero equal to maltodextrin	[75]
$n = 31$, 67.4 ± 4.0 years, females 8-week pre-conditioned resistance training Randomised, double-blind controlled	12 weeks 35 g hydrolysed whey protein (Lacprodan, Arla Foods) Exercise = chest press, knee extension and preacher curl Whey $n = 13$	↑ Chest press, knee extension, preacher curl, total strength, lean soft tissue, muscle mass, muscle quality index V 8-week exercise alone ↑ Chest press, knee extension total strength V maltodextrin with exercise	[77]
* $n = 75$, aged 65–80 years, females Randomised, single-blind controlled.	24 weeks 22.3 g whey protein (Ezaki Glico) Exercise = rising and sitting from a chair, and leg extension with resistance band exercises Whey $n = 54$	Exercise + whey > exercise > whey for ↑ limb muscle mass, skeletal muscle mass index, grip strength, knee extension, gait speed	[78]
* $n = 149$ aged 78.5 ± 5.4 years, males and females Randomised, double blinded, placebo-controlled	6-month period whey protein 20 g + Vitamin D 800 IU (Nestle Health Science) Exercise = 30 min aerobic + 20 min strength Whey $n = 74$	↑ Normal density muscle V controls (a low-calorie placebo drink -30 kcal, no protein, no Vitamin D) and time zero ↓ Low-density muscle V time zero ↓ Intermuscular fat V controls and time zero ↑ Knee flexor strength, power and quality for both supplemented and control groups V baseline values	[79,80]

Table 4. *Cont.*

Cohort	Whey Source, Dosage, Duration	Outcome of Whey Supplementation	Reference
* n = 80 aged 70–85 years, males and females Randomised, double-blind, controlled	6 months Whey protein concentrate 40 g/daily (Innovative food processors Inc.) Exercise = walking or stationary cycling, knee extension and leg press Whey n = 42	↑ Knee extensor power, total lean mass, total mid-thigh CSA, total muscle CSA and total normal density muscle CSA V maltodextrin ↑ Peak power for double leg press V time zero, but similar to maltodextrin ↓ Stair climb and chair rise time V time zero	[81]
* n = 46 aged ≥60 – 86 years, males and females Randomised, single-blind	16 weeks Leucine (0.03 g/kg/meal) enriched whey protein (0.5 g/kg/day) (MyProtein) Exercise = leg press, chest press and bicep curl Whey n = 22	↑ Leg press, chest press and bicep curl V time zero, no differences between whey + exercise or exercise groups ↑ SPPB score V time zero	[82]
* n = 100 aged 69 ± 6 years, males and females Randomised, single-blind, controlled	16 weeks Leucine (0.03 g/kg/meal) enriched whey protein (0.5 g/kg/day) (MyProtein) Exercise = leg press, chest press, calf press, shoulder press, seated row, back extension and bicep curl Whey n = 45	Rectus femoris and bicep femoris were more resistant to fatigue for both exercise intervention V time zero No additional benefit with supplementation of whey	[83]
n = 65, aged = 60–93 years, hospitalised males and females Single-blind test, controlled	Length of hospital stay 24 g of whey protein Rehabilitation programme Whey n = 27	↑ Grip strength, knee extensor force V time zero improved gait speed, and timed up and go V time zero ↓ IL-6 V time zero for both WPS group & control group (hot cereal and milk products - WPS) >↓ in % change of IL-6 V controls (hot cereal and milk products - WPS)	[84]
* n = 165 aged >70 years, males and females hospitalised subjects Block randomised, double-blinded, placebo-controlled	Length of hospitalization and 12 weeks after discharge A protein-enriched, milk-based supplement beverage–26.25 g whey protein (Protino, Arla foods) Exercise = bridging exercise, rising from a chair, and heel-raise Whey n = 83	No noted benefit of whey protein supplementation Iso-energetic placebo-products (<1.5 g protein) = improved 30 sec chair stand test, 4 m gait speed, ↑ handgrip strength V time zero	[85,86]

Table 5. Older adult intervention with whey protein supplementation but no exercise. IGF-1: Insulin growth factor 1; EAAMR: Essential amino acid meal replacement; FSR: fractional synthesis rate. * power calculation to determine number of subjects was performed.

Cohort	Whey Source, Dosage, Duration	Outcome of Whey Supplementation	Reference
n = 69 aged 52.5 ± 15.2 years, males and females. Double-blind, cross-over study	30 days Whey based oral supplement-10 g/day (ProtherSOD) Exercise = cycle ergometer Whey n = 13	↑ FRAP and GSH levels V time zero and casein ↓ AOPP V time zero and casein	[87]
* n = 24 aged 71 ± 4 years, males Randomised, double-blind, placebo-controlled	6 weeks Leucine (3 g) enriched whey protein (20 g) in combination with vitamin D (800 IU) (Nutricia Advanced Medical Nutrition) Whey n = 12	↑ Mixed muscle protein synthesis rate V time zero and control (flavored water drink). Controls = ↑ V time zero ↑ Appendicular muscle mass and leg lean mass V flavored water	[88]
n = 52 aged 85.0 ± 7.4 years, males and females Randomised, double-blind, controlled	28-day intervention 15 g whey protein + 5 g essential amino acids + 550 mg Ca (Novartis Nutrition) Whey n = 61	↑ Activity of daily living score for whey ± zinc V time zero ↑ IGF-1 in whey ± zinc V time zero	[89,90]
n = 12 aged 65–80 years, males and females Randomised	8 weeks EAAMR: Essential amino acid meal replacement–7 g intact whey protein Whey n = 6	↑ Skeletal muscle FSR V control group (competitive meal replacer)	[91]

Studies describing whey supplementation with whey protein on sarcopenic individuals are detailed in Table 6. In longer term studies, Nabuco et al. [92] investigated the effects of whey in 26 elderly women > 60 years of age, with sarcopenic obesity, who consumed 35 g hydrolysed whey protein over a 16-week period in tandem with resistance training (chest press, horizontal leg press, seated row, knee extension, preacher curl, leg curl, triceps pushdown, seated calf raise). Whey protein significantly reduced circulating IL-6 compared to maltodextrin and time zero. However, the vast majority of sarcopenic biomarkers were improved by exercise rather than whey. For instance, at the end of the trial, there was a significant reduction in the number of individuals defined as sarcopenic, and a significant decrease in circulating TNF-α and CRP with an increase in blood antioxidant potential compared to time zero but this was similar for whey or maltodextrin. Liberman et al. [93] investigated the effects of a 13-week supplementation programme of vitamin D (800 IU) and leucine-enriched whey protein (3 g leucine/20 g WP) on the chronic low-grade inflammatory profile (CLIP) in sarcopenic people (aged $\geq$ 65 years n = 297). For plasma IL-6, a treatment x time interaction was noted, with no significant difference over time for the whey group compared to a significant increase for the control group (31.4 g of carbohydrates). There was no significant difference in plasma IL-1Ra or IL-8 following whey protein supplementation. Bo et al. [94] investigated the effects of whey protein supplementation (57.5% protein) in combination with vitamin D (702 IU) and E (109 mg) on sarcopenic individuals aged 60–85 years old (n = 60). After 6 months of whey protein supplementation, appendicular muscle mass, relative skeletal muscle mass index, handgrip strength and serum IGF-1 were all increased compared to the isocaloric placebo (32.4 g of carbohydrates, 2.6 g of fat). In addition, there was a significant decrease in serum IL-2 and in the time taken to perform a 'time to stand test' in comparison to the placebo group. Li et al. [95] compared the supplementation of whey protein in tandem with resistance training in sarcopenic individuals (72.05 ± 6.54 years old) to non-sarcopenic individuals (65.24 ± 4.05) (n = 56 per group) over a 12-week period. The whey protein supplement contained 30 g of whey protein and 3.84 g of leucine per serving taken twice daily. Following whey protein intervention in sarcopenic individuals, there was a significant increase in muscle mass and IGF-1 compared to time zero. There was a significant decrease in TNF-like weak inducer of apoptosis (TWEAK), TNF-α and IL-18 compared to pre-intervention parameters [95]. The PROVIDE study carried out by Hill et al. [96] aimed to investigate the effects of leucine (3 g) enriched whey protein (20 g) supplementation in elderly sarcopenic individuals aged > 65 years (n = 380) over a 13-week period. Serum IGF-1 was significantly increased in the whey protein supplementation group compared to the isocaloric control. For the same cohort of individuals, Bauer et al. [97] noted that appendicular muscle mass, hand grip strength, chair stand test time, short physical performance battery score and gait speed were significantly improved in the whey protein group compared to time zero. However, the improvement in several of these biomarkers (hand grip strength, gait speed, SPPB, chair-stand time and balance test) were similar to that observed in the isocaloric control group [96,97].

Although the trials used different scores to classify sarcopenic individuals [95,96], overall whey supplementation improved sarcopenic biomarkers, with a somewhat clearer outcome emerging than data from healthy individuals of similar age. The majority of studies in human trials lasted 4 months [92], with a dosage range of 20–35 g [93–95] whey protein per day. There were a number of notable limitations to the human studies. First and foremost was the omission of power calculations in the description of experimental design in 14 trials. In addition, a number of studies including Niccoli et al. [84], Mancuso et al. [87], Reitelseder et al. [68], and Borack et al. [69] had whey in their control group, which confounded the results and made interpretation difficult. Twelve studies included a combination of whey with exercise, making it difficult to unravel the contribution of whey alone without appropriate controls. Many studies observed significant differences to time zero instead of isocaloric [85,96], maltodextrin [75–77] or matched protein [68,87] control groups. In addition, many of the studies employed different whey products or ingredients which also made comparisons difficult.

Table 6. Whey intervention trials involving older individuals with sarcopenia. LST: Lean soft tissue; IL-6: Interleukin 6; TNF-α: Tumor necrosis factor alpha; CRP: C reactive protein; BIA: Bio-Impedance Analysis; IL-1Ra: Interleukin 1 receptor antagonist; IL-8: Interleukin 8; RSMI: Relative muscle mass index; IGF-1: Insulin-like growth factor 1; IL-2: Interleukin 2; TWEAK—Tumor necrosis factor-like WEAK inducer of apoptosis; IL-18 Interleukin 18; SPPB: Short Physical Performance Battery. * power calculation to determine number of subjects was performed.

Cohort	Whey Source, Dosage, Duration	Outcome of Whey Supplementation	Reference
$n = 26 \geq 60$ years, females sarcopenic obesity—fat mass ≥35% & appendicular lean soft tissue ≤15.02 kg Randomised, double blind, placebo controlled	16 weeks 35 g hydrolysed whey protein (Lacprodan, Arla Foods) Exercise = chest press, horizontal leg press, seated row, knee extension, preacher curl, leg curl, triceps pushdown, and seated calf raise Whey $n = 13$	↑ LST V maltodextrin and time zero ↓ Total LST, relative total fat mass and trunk fat mass V maltodextrin ↓ Number of sarcopenia and sarcopenic obesity V time zero ↓ Blood IL-6 V time zero and maltodextrin ↑ Knee extension, chest press, preacher curl, total strength V time zero but no difference to maltodextrin ↓ Blood TNF-a, CRP, advanced oxidation products V time zero but not maltodextrin ↑ Blood total radical trapping antioxidant potential V time zero but not maltodextrin	[92]
* $n = 297$ aged ≥ 65 years, males and females Sarcopenic subjects: BIA < 37% in men and <28% women Randomised double blind isocaloric controlled.	13 weeks 20 g whey protein, 3 g total leucine, and 800 IU vitamin D Whey $n = 137$	↑ IL-6, IL-1Ra & CRP for control (31.4 g of carbohydrates) and supplemented groups V time zero ↓ IL-8 V time zero	[93]
$n = 60$ aged 60–85 years, males and females Sarcopenic subject: RSMI < 5.7 kg/m^2 for women and <7.0 kg/m^2 for men Double blind randomized controlled	6-month 22 g whey protein plus vitamin D (702 IU) & E (109 mg) per serving Whey $n = 30$	↑ Appendicular muscle mass, RSMI, handgrip strength, IGF-1 V isocaloric placebo (32.4 g of carbohydrates, 2.6 g of fat) ↓ IL-2 V isocaloric placebo No difference TNF-a, IL-6, CRP V isocaloric placebo or time zero ↓ Time to stand V isocaloric placebo	[94]
* $n = 112$ males & females, n = 56 sarcopenic aged 72.05 ± 6.54, n = 56 non sarcopenic aged 65.24 ± 4.05 Sarcopenia subjects: RSMI 7.0 kg/m^2 men, 5.4 kg/m^2 women	12 weeks 30 g of whey with 3.84 g total leucine. (Lacprodan, Arla Foods) Exercise = 5 min warm-up, 20 min muscle strength training, and 5 min slow walking	Sarcopenic group: ↑ muscle mass V time zero ↓ TWERK, TNF-α and IL-18 V time zero. ↑ IGF-1 V time zero	[95]
* $n = 380$ aged > 65 years, males & females Sarcopenia = SPPB score between 4–9 and a skeletal muscle index of ≤37% in men and ≤28% in women Randomised, double-blind, controlled	13 weeks Leucine (3 g) enriched whey protein (20 g) Whey $n = 184$	↑ IGF-1, V time zero.↓ in IGF-1 for control group (isocaloric product) V time zero ↓ Chair stand test time and gait speed V time zero for both test and control groups ↑ SPPB score V time zero for both test and isocaloric product	[96] [97]

5. Conclusions

In summary, a daily dietary supplementation of 35 g whey is likely to improve sarcopenic biomarkers in frail or sarcopenic individuals. For the older healthy adult, whey supplementation certainly improves muscle mTOR signaling, but the greatest benefit to the older muscle is exercise. In vitro cellular assays will continue to play an essential role in identifying bioactive and bioavailable peptides from whey and in unraveling their mechanism of action on the ageing muscle. Future studies should also investigate emerging biomarkers to assess association with sarcopenia, e.g., isoprostanes [98,99] or allantoin [100] as indicators of oxidative stress.

Author Contributions: S.G.; writing–original draft preparation, L.G.; data curation, manuscript reviewing and, funding acquisition and supervision. N.O.; manuscript review and editing; funding acquisition and supervision. All authors have read and agreed to the published version of the manuscript.

Funding: This work was supported by Teagasc (MDBY0015-PRO4FOOD) the Irish Department of Agriculture, Food and the Marine (FIRM 15F604-TOMI) and Science Foundation Ireland with Irish Department of Agriculture Food and the Marine (16/RC/3835-VistaMilk). S.G. is in receipt of a Teagasc Walsh Scholarship.

Conflicts of Interest: The authors declare no conflict of interest.

References

1. Tieland, M.; Trouwborst, I.; Clark, B.C. Skeletal muscle performance and ageing. *J. Cachexiasarcopenia Muscle* **2018**, *9*, 3–19. [CrossRef] [PubMed]

2. Aagaard, P.; Suetta, C.; Caserotti, P.; Magnusson, S.P.; Kjaer, M. Role of the nervous system in sarcopenia and muscle atrophy with aging: Strength training as a countermeasure. *Scand. J. Med. Sci. Sports* **2010**, *20*, 49–64. [CrossRef] [PubMed]

3. Kalyani, R.R.; Corriere, M.; Ferrucci, L. Age-related and disease-related muscle loss: The effect of diabetes, obesity, and other diseases. *Lancet Diabetes Endocrinol.* **2014**, *2*, 819–829. [CrossRef]

4. Patel, H.P.; Syddall, H.E.; Jameson, K.; Robinson, S.; Denison, H.; Roberts, H.C.; Edwards, M.; Dennison, E.; Cooper, C.; Aihie Sayer, A. Prevalence of sarcopenia in community-dwelling older people in the UK using the European Working Group on Sarcopenia in Older People (EWGSOP) definition: Findings from the Hertfordshire Cohort Study (HCS). *Age Ageing* **2013**, *42*, 378–384. [CrossRef]

5. MacEwan, J.P.; Gill, T.M.; Johnson, K.; Doctor, J.; Sullivan, J.; Shim, J.; Goldman, D.P. Measuring Sarcopenia Severity in Older Adults and the Value of Effective Interventions. *J. Nutr. Health Aging* **2018**, *22*, 1253–1258. [CrossRef] [PubMed]

6. Von Haehling, S.; Morley, J.E.; Anker, S.D. An overview of sarcopenia: Facts and numbers on prevalence and clinical impact. *J. Cachexiasarcopenia Muscle* **2010**, *1*, 129–133. [CrossRef] [PubMed]

7. Ethgen, O.; Beaudart, C.; Buckinx, F.; Bruyere, O.; Reginster, J.Y. The Future Prevalence of Sarcopenia in Europe: A Claim for Public Health Action. *Calcif. Tissue Int.* **2017**, *100*, 229–234. [CrossRef] [PubMed]

8. Shafiee, G.; Keshtkar, A.; Soltani, A.; Ahadi, Z.; Larijani, B.; Heshmat, R. Prevalence of sarcopenia in the world: A systematic review and meta- analysis of general population studies. *J. Diabetes Metab. Disord.* **2017**, *16*, 21. [CrossRef] [PubMed]

9. Cruz-Jentoft, A.J.; Bahat, G.; Bauer, J.; Boirie, Y.; Bruyère, O.; Cederholm, T.; Cooper, C.; Landi, F.; Rolland, Y.; Sayer, A.A.; et al. Sarcopenia: Revised European consensus on definition and diagnosis. *Age Ageing* **2019**, *48*, 16–31. [CrossRef]

10. Chumlea, W.C.; Cesari, M.; Evans, W.J.; Ferrucci, L.; Fielding, R.A.; Pahor, M.; Studenski, S.; Vellas, B.; The Task Force, M. International working group on Sarcopenia. *J. Nutr. Health Aging* **2011**, *15*, 450–455. [CrossRef]

11. Chen, L.K.; Woo, J.; Assantachai, P.; Auyeung, T.W.; Chou, M.Y.; Iijima, K.; Jang, H.C.; Kang, L.; Kim, M.; Kim, S.; et al. Asian Working Group for Sarcopenia: 2019 Consensus Update on Sarcopenia Diagnosis and Treatment. *J. Am. Med. Dir. Assoc.* **2020**, *21*, 300–307.e302. [CrossRef] [PubMed]

12. Messina, C.; Maffi, G.; Vitale, J.A.; Ulivieri, F.M.; Guglielmi, G.; Sconfienza, L.M. Diagnostic imaging of osteoporosis and sarcopenia: A narrative review. *Quant. Imaging Med. Surg.* **2018**, *8*, 86–99. [CrossRef] [PubMed]

13. Kim, J.; Wang, Z.; Heymsfield, S.B.; Baumgartner, R.N.; Gallagher, D. Total-body skeletal muscle mass: Estimation by a new dual-energy X-ray absorptiometry method. *Am. J. Clin. Nutr.* **2002**, *76*, 378–383. [CrossRef] [PubMed]

14. Guimarães-Ferreira, L. Role of the phosphocreatine system on energetic homeostasis in skeletal and cardiac muscles. *Einstein (Sao Paulo, Brazil)* **2014**, *12*, 126–131. [CrossRef]

15. Stevens, L.A.; Coresh, J.; Greene, T.; Levey, A.S. Assessing kidney function–measured and estimated glomerular filtration rate. *N. Engl. J. Med.* **2006**, *354*, 2473–2483. [CrossRef]

16. Patel, S.S.; Molnar, M.Z.; Tayek, J.A.; Ix, J.H.; Noori, N.; Benner, D.; Heymsfield, S.; Kopple, J.D.; Kovesdy, C.P.; Kalantar-Zadeh, K. Serum creatinine as a marker of muscle mass in chronic kidney disease: Results of a cross-sectional study and review of literature. *J. Cachexiasarcopenia Muscle* **2013**, *4*, 19–29. [CrossRef]

17. Rong, Y.-D.; Bian, A.-L.; Hu, H.-Y.; Ma, Y.; Zhou, X.-Z. Study on relationship between elderly sarcopenia and inflammatory cytokine IL-6, anti-inflammatory cytokine IL-10. *BMC Geriatr.* **2018**, *18*, 308. [CrossRef]

18. Bian, A.-L.; Hu, H.-Y.; Rong, Y.-D.; Wang, J.; Wang, J.-X.; Zhou, X.-Z. A study on relationship between elderly sarcopenia and inflammatory factors IL-6 and TNF-α. *Eur. J. Med Res.* **2017**, *22*, 25. [CrossRef]

19. Osaka, T.; Hamaguchi, M.; Hashimoto, Y.; Ushigome, E.; Tanaka, M.; Yamazaki, M.; Fukui, M. Decreased the creatinine to cystatin C ratio is a surrogate marker of sarcopenia in patients with type 2 diabetes. *Diabetes Res. Clin. Pr.* **2018**, *139*, 52–58. [CrossRef]

20. Ferrucci, L.; Fabbri, E. Inflammageing: Chronic inflammation in ageing, cardiovascular disease, and frailty. *Nat. Rev. Cardiol.* **2018**, *15*, 505–522. [CrossRef]

21. Kwak, J.Y.; Hwang, H.; Kim, S.-K.; Choi, J.Y.; Lee, S.-M.; Bang, H.; Kwon, E.-S.; Lee, K.-P.; Chung, S.G.; Kwon, K.-S. Prediction of sarcopenia using a combination of multiple serum biomarkers. *Sci. Rep.* **2018**, *8*, 8574. [CrossRef] [PubMed]

22. Taaffe, D.R.; Harris, T.B.; Ferrucci, L.; Rowe, J.; Seeman, T.E. Cross-sectional and prospective relationships of interleukin-6 and C-reactive protein with physical performance in elderly persons: MacArthur studies of successful aging. *J. Gerontol. Ser. Abiological Sci. Med Sci.* **2000**, *55*, M709–M715. [CrossRef] [PubMed]

23. Bano, G.; Trevisan, C.; Carraro, S.; Solmi, M.; Luchini, C.; Stubbs, B.; Manzato, E.; Sergi, G.; Veronese, N. Inflammation and sarcopenia: A systematic review and meta-analysis. *Maturitas* **2017**, *96*, 10–15. [CrossRef] [PubMed]

24. Stowe, R.P.; Peek, M.K.; Goodwin, J.S.; Cutchin, M.P. Plasma Cytokine Levels in a Population-Based Study: Relation to Age and Ethnicity. *J. Gerontol. Ser. A* **2009**, *65A*, 429–433. [CrossRef] [PubMed]

25. Leger, B.; Derave, W.; De Bock, K.; Hespel, P.; Russell, A.P. Human sarcopenia reveals an increase in SOCS-3 and myostatin and a reduced efficiency of Akt phosphorylation. *Rejuvenation Res.* **2008**, *11*, 163–175b. [CrossRef] [PubMed]

26. Barclay, R.D.; Burd, N.A.; Tyler, C.; Tillin, N.A.; MacKenzie, R.W. The Role of the IGF-1 Signaling Cascade in Muscle Protein Synthesis and Anabolic Resistance in Aging Skeletal Muscle. *Front. Nutr.* **2019**, *6*, 146. [CrossRef] [PubMed]

27. Ashpole, N.M.; Herron, J.C.; Estep, P.N.; Logan, S.; Hodges, E.L.; Yabluchanskiy, A.; Humphrey, M.B.; Sonntag, W.E. Differential effects of IGF-1 deficiency during the life span on structural and biomechanical properties in the tibia of aged mice. *Age (Dordr)* **2016**, *38*, 38. [CrossRef]

28. Ng, T.P.; Lu, Y.; Choo, R.W.M.; Tan, C.T.Y.; Nyunt, M.S.Z.; Gao, Q.; Mok, E.W.H.; Larbi, A. Dysregulated homeostatic pathways in sarcopenia among frail older adults. *Aging Cell* **2018**, *17*, e12842. [CrossRef]

29. Pansarasa, O.; Castagna, L.; Colombi, B.; Vecchiet, J.; Felzani, G.; Marzatico, F. Age and sex differences in human skeletal muscle: Role of reactive oxygen species. *Free Radic. Res.* **2000**, *33*, 287–293. [CrossRef]

30. Young, I.S.; Woodside, J.V. Antioxidants in health and disease. *J. Clin. Pathol.* **2001**, *54*, 176–186. [CrossRef]

31. Powers, S.K.; Ji, L.L.; Kavazis, A.N.; Jackson, M.J. Reactive oxygen species: Impact on skeletal muscle. *Compr. Physiol.* **2011**, *1*, 941–969. [PubMed]

32. Powers, S.K.; Jackson, M.J. Exercise-induced oxidative stress: Cellular mechanisms and impact on muscle force production. *Physiol. Rev.* **2008**, *88*, 1243–1276. [CrossRef] [PubMed]

33. Kurutas, E.B. The importance of antioxidants which play the role in cellular response against oxidative/nitrosative stress: Current state. *Nutr. J.* **2016**, *15*, 71. [CrossRef] [PubMed]

34. Liguori, I.; Russo, G.; Curcio, F.; Bulli, G.; Aran, L.; Della-Morte, D.; Gargiulo, G.; Testa, G.; Cacciatore, F.; Bonaduce, D.; et al. Oxidative stress, aging, and diseases. *Clin. Interv. Aging* **2018**, *13*, 757–772. [CrossRef] [PubMed]

35. Frontera, W.R.; Ochala, J. Skeletal muscle: A brief review of structure and function. *Calcif. Tissue Int.* **2015**, *96*, 183–195. [CrossRef]

36. Liao, Y.; Peng, Z.; Chen, L.; Zhang, Y.; Cheng, Q.; Nüssler, A.K.; Bao, W.; Liu, L.; Yang, W. Prospective Views for Whey Protein and/or Resistance Training Against Age-related Sarcopenia. *Aging Dis* **2019**, *10*, 157–173. [CrossRef]

37. West, D.W.D.; Abou Sawan, S.; Mazzulla, M.; Williamson, E.; Moore, D.R. Whey Protein Supplementation Enhances Whole Body Protein Metabolism and Performance Recovery after Resistance Exercise: A Double-Blind Crossover Study. *Nutrients* **2017**, *9*, 735. [CrossRef]

38. Davies, R.W.; Carson, B.P.; Jakeman, P.M. The Effect of Whey Protein Supplementation on the Temporal Recovery of Muscle Function Following Resistance Training: A Systematic Review and Meta-Analysis. *Nutrients* **2018**, *10*, 221. [CrossRef]

39. Stark, M.; Lukaszuk, J.; Prawitz, A.; Salacinski, A. Protein timing and its effects on muscular hypertrophy and strength in individuals engaged in weight-training. *J. Int. Soc. Sports Nutr.* **2012**, *9*, 54. [CrossRef]

40. Rutherfurd, S.M.; Fanning, A.C.; Miller, B.J.; Moughan, P.J. Protein Digestibility-Corrected Amino Acid Scores and Digestible Indispensable Amino Acid Scores Differentially Describe Protein Quality in Growing Male Rats. *J. Nutr.* **2014**, *145*, 372–379. [CrossRef]

41. Brennan, J.L.; Keerati-U-Rai, M.; Yin, H.; Daoust, J.; Nonnotte, E.; Quinquis, L.; St-Denis, T.; Bolster, D.R. Differential Responses of Blood Essential Amino Acid Levels Following Ingestion of High-Quality Plant-Based Protein Blends Compared to Whey Protein-A Double-Blind Randomized, Cross-Over, Clinical Trial. *Nutrients* **2019**, *11*, 2987. [CrossRef] [PubMed]

42. Almeida, C.C.; Alvares, T.S.; Costa, M.P.; Conte-Junior, C.A. Protein and Amino Acid Profiles of Different Whey Protein Supplements. *J. Diet. Suppl.* **2016**, *13*, 313–323. [CrossRef] [PubMed]

43. Hulmi, J.J.; Lockwood, C.M.; Stout, J.R. Effect of protein/essential amino acids and resistance training on skeletal muscle hypertrophy: A case for whey protein. *Nutr. Metab. (Lond.)* **2010**, *7*, 51. [CrossRef] [PubMed]

44. Tosukhowong, P.; Boonla, C.; Dissayabutra, T.; Kaewwilai, L.; Muensri, S.; Chotipanich, C.; Joutsa, J.; Rinne, J.; Bhidayasiri, R. Biochemical and clinical effects of Whey protein supplementation in Parkinson's disease: A pilot study. *J. Neurol. Sci.* **2016**, *367*, 162–170. [CrossRef] [PubMed]

45. Hulmi, J.J.; Tannerstedt, J.; Selanne, H.; Kainulainen, H.; Kovanen, V.; Mero, A.A. Resistance exercise with whey protein ingestion affects mTOR signaling pathway and myostatin in men. *J. Appl. Physiol. (Bethesda Md. 1985)* **2009**, *106*, 1720–1729. [CrossRef] [PubMed]

46. Brandelli, A.; Daroit, D.J.; Corrêa, A.P.F. Whey as a source of peptides with remarkable biological activities. *Food Res. Int.* **2015**, *73*, 149–161. [CrossRef]

47. Madureira, A.R.; Pereira, C.I.; Gomes, A.M.P.; Pintado, M.E.; Xavier Malcata, F. Bovine whey proteins–Overview on their main biological properties. *Food Res. Int.* **2007**, *40*, 1197–1211. [CrossRef]

48. Adams, R.L.; Broughton, K.S. Insulinotropic Effects of Whey: Mechanisms of Action, Recent Clinical Trials, and Clinical Applications. *Ann. Nutr. Metab.* **2016**, *69*, 56–63. [CrossRef]

49. Corrochano, A.R.; Ferraretto, A.; Arranz, E.; Stuknyte, M.; Bottani, M.; O'Connor, P.M.; Kelly, P.M.; De Noni, I.; Buckin, V.; Giblin, L. Bovine whey peptides transit the intestinal barrier to reduce oxidative stress in muscle cells. *Food Chem.* **2019**, *288*, 306–314. [CrossRef]

50. Ogiwara, M.; Ota, W.; Mizushige, T.; Kanamoto, R.; Ohinata, K. Enzymatic digest of whey protein and wheylin-1, a dipeptide released in the digest, increase insulin sensitivity in an Akt phosphorylation-dependent manner. *Food Funct.* **2018**, *9*, 4635–4641. [CrossRef]

51. Mobley, C.B.; Fox, C.D.; Thompson, R.M.; Healy, J.C.; Santucci, V.; Kephart, W.C.; McCloskey, A.E.; Kim, M.; Pascoe, D.D.; Martin, J.S.; et al. Comparative effects of whey protein versus l-leucine on skeletal muscle protein synthesis and markers of ribosome biogenesis following resistance exercise. *Amino Acids* **2016**, *48*, 733–750. [CrossRef] [PubMed]

52. Frey, J.W.; Jacobs, B.L.; Goodman, C.A.; Hornberger, T.A. A role for Raptor phosphorylation in the mechanical activation of mTOR signaling. *Cell Signal* **2014**, *26*, 313–322. [CrossRef]

53. Kerasioti, E.; Stagos, D.; Priftis, A.; Aivazidis, S.; Tsatsakis, A.M.; Hayes, A.W.; Kouretas, D. Antioxidant effects of whey protein on muscle C2C12 cells. *Food Chem.* **2014**, *155*, 271–278. [CrossRef]

54. Xu, R.; Liu, N.; Xu, X.; Kong, B. Antioxidative effects of whey protein on peroxide-induced cytotoxicity. *J. Dairy Sci.* **2011**, *94*, 3739–3746. [CrossRef] [PubMed]

55. Knight, M.I.; Tester, A.M.; McDonagh, M.B.; Brown, A.; Cottrell, J.; Wang, J.; Hobman, P.; Cocks, B.G. Milk-derived ribonuclease 5 preparations induce myogenic differentiation in vitro and muscle growth in vivo. *J. Dairy Sci.* **2014**, *97*, 7325–7333. [CrossRef] [PubMed]

56. Carson, B.P.; Patel, B.; Amigo-Benavent, M.; Pauk, M.; Kumar Gujulla, S.; Murphy, S.M.; Kiely, P.A.; Jakeman, P.M. Regulation of muscle protein synthesis in an in vitro cell model using ex vivo human serum. *Exp. Physiol.* **2018**, *103*, 783–789. [CrossRef] [PubMed]

57. Dutta, S.; Sengupta, P. Men and mice: Relating their ages. *Life Sci.* **2016**, *152*, 244–248. [CrossRef]

58. Andreollo, N.A.; Santos, E.F.d.; Araújo, M.R.; Lopes, L.R. Idade dos ratos versus idade humana: Qual é a relação? *Abcd. Arq. Bras. De Cir. Dig. (São Paulo)* **2012**, *25*, 49–51. [CrossRef]

59. Van Dijk, M.; Dijk, F.J.; Bunschoten, A.; van Dartel, D.A.; van Norren, K.; Walrand, S.; Jourdan, M.; Verlaan, S.; Luiking, Y. Improved muscle function and quality after diet intervention with leucine-enriched whey and antioxidants in antioxidant deficient aged mice. *Oncotarget* **2016**, *7*, 17338–17355. [CrossRef]

60. Mosoni, L.; Gatineau, E.; Gatellier, P.; Migne, C.; Savary-Auzeloux, I.; Remond, D.; Rocher, E.; Dardevet, D. High whey protein intake delayed the loss of lean body mass in healthy old rats, whereas protein type and polyphenol/antioxidant supplementation had no effects. *PLoS ONE* **2014**, *9*, e109098. [CrossRef]

61. Garg, G.; Singh, S.; Kumar Singh, A.; Ibrahim Rizvi, S. Whey protein concentrate supplementation protects erythrocyte membrane from aging-induced alterations in rats. *J. Food Biochem.* **2018**, *42*, e12679-n/a. [CrossRef]

62. Dijk, F.J.; van Dijk, M.; Walrand, S.; van Loon, L.J.C.; van Norren, K.; Luiking, Y.C. Differential effects of leucine and leucine-enriched whey protein on skeletal muscle protein synthesis in aged mice. *Clin. Nutr. ESPEN* **2018**, *24*, 127–133. [CrossRef] [PubMed]

63. Jarzaguet, M.; Polakof, S.; David, J.; Migne, C.; Joubrel, G.; Efstathiou, T.; Remond, D.; Mosoni, L.; Dardevet, D. A meal with mixed soy/whey proteins is as efficient as a whey meal in counteracting the age-related muscle anabolic resistance only if the protein content and leucine levels are increased. *Food Funct.* **2018**, *9*, 6526–6534. [CrossRef] [PubMed]

64. Walrand, S.; Zangarelli, A.; Guillet, C.; Salles, J.; Soulier, K.; Giraudet, C.; Patrac, V.; Boirie, Y. Effect of fast dietary proteins on muscle protein synthesis rate and muscle strength in ad libitum-fed and energy-restricted old rats. *Br. J. Nutr.* **2011**, *106*, 1683–1690. [CrossRef] [PubMed]

65. Lafoux, A.; Baudry, C.; Bonhomme, C.; Le Ruyet, P.; Huchet, C. Soluble Milk Protein Supplementation with Moderate Physical Activity Improves Locomotion Function in Aging Rats. *PLoS ONE* **2016**, *11*, e0167707. [CrossRef]

66. Magne, H.; Savary-Auzeloux, I.; Migné, C.; Peyron, M.-A.; Combaret, L.; Rémond, D.; Dardevet, D. Contrarily to whey and high protein diets, dietary free leucine supplementation cannot reverse the lack of recovery of muscle mass after prolonged immobilization during ageing. *J. Physiol.* **2012**, *590*, 2035–2049. [CrossRef]

67. Pennings, B.; Groen, B.; Lange, A.d.; Gijsen, A.P.; Zorenc, A.H.; Senden, J.M.G.; Loon, L.J.C.v. Amino acid absorption and subsequent muscle protein accretion following graded intakes of whey protein in elderly men. *Am. J. Physiol. Endocrinol. Metab.* **2012**, *302*, E992–E999. [CrossRef]

68. Reitelseder, S.; Dideriksen, K.; Agergaard, J.; Malmgaard-Clausen, N.M.; Bechshoeft, R.L.; Petersen, R.K.; Serena, A.; Mikkelsen, U.R.; Holm, L. Even effect of milk protein and carbohydrate intake but no further effect of heavy resistance exercise on myofibrillar protein synthesis in older men. *Eur. J. Nutr.* **2019**, *58*, 583–595. [CrossRef]

69. Borack, M.S.; Reidy, P.T.; Husaini, S.H.; Markofski, M.M.; Deer, R.R.; Richison, A.B.; Lambert, B.S.; Cope, M.B.; Mukherjea, R.; Jennings, K.; et al. Soy-Dairy Protein Blend or Whey Protein Isolate Ingestion Induces Similar Postexercise Muscle Mechanistic Target of Rapamycin Complex 1 Signaling and Protein Synthesis Responses in Older Men. *J. Nutr.* **2016**, *146*, 2468–2475. [CrossRef]

70. Wilkinson, D.J.; Bukhari, S.S.I.; Phillips, B.E.; Limb, M.C.; Cegielski, J.; Brook, M.S.; Rankin, D.; Mitchell, W.K.; Kobayashi, H.; Williams, J.P.; et al. Effects of leucine-enriched essential amino acid and whey protein bolus dosing upon skeletal muscle protein synthesis at rest and after exercise in older women. *Clin. Nutr. (Edinb. Scotl.)* **2018**, *37*, 2011–2021. [CrossRef]

71. Kramer, I.F.; Verdijk, L.B.; Hamer, H.M.; Verlaan, S.; Luiking, Y.C.; Kouw, I.W.K.; Senden, J.M.; van Kranenburg, J.; Gijsen, A.P.; Bierau, J.; et al. Both basal and post-prandial muscle protein synthesis rates, following the ingestion of a leucine-enriched whey protein supplement, are not impaired in sarcopenic older males. *Clin. Nutr. (Edinb. Scotl.)* **2017**, *36*, 1440–1449. [CrossRef] [PubMed]

72. Dideriksen, K.; Reitelseder, S.; Malmgaard-Clausen, N.M.; Bechshoeft, R.; Petersen, R.K.; Mikkelsen, U.R.; Holm, L. No effect of anti-inflammatory medication on postprandial and postexercise muscle protein synthesis in elderly men with slightly elevated systemic inflammation. *Exp. Gerontol.* **2016**, *83*, 120–129. [CrossRef] [PubMed]

73. Smith, G.I.; Yoshino, J.; Stromsdorfer, K.L.; Klein, S.J.; Magkos, F.; Reeds, D.N.; Klein, S.; Mittendorfer, B. Protein Ingestion Induces Muscle Insulin Resistance Independent of Leucine-Mediated mTOR Activation. *Diabetes* **2015**, *64*, 1555–1563. [CrossRef] [PubMed]

74. Mayer, F.; Scharhag-Rosenberger, F.; Carlsohn, A.; Cassel, M.; Müller, S.; Scharhag, J. The intensity and effects of strength training in the elderly. *Dtsch. Arztebl. Int.* **2011**, *108*, 359–364. [CrossRef] [PubMed]

75. Nabuco, H.C.G.; Tomeleri, C.M.; Sugihara Junior, P.; Fernandes, R.R.; Cavalcante, E.F.; Antunes, M.; Ribeiro, A.S.; Teixeira, D.C.; Silva, A.M.; Sardinha, L.B.; et al. Effects of Whey Protein Supplementation Pre- or Post-Resistance Training on Muscle Mass, Muscular Strength, and Functional Capacity in Pre-Conditioned Older Women: A Randomized Clinical Trial. *Nutrients* **2018**, *10*, 563. [CrossRef] [PubMed]

76. Nabuco, H.C.G.; Tomeleri, C.M.; Fernandes, R.R.; Sugihara Junior, P.; Venturini, D.; Barbosa, D.S.; Deminice, R.; Sardinha, L.B.; Cyrino, E.S. Effects of pre- or post-exercise whey protein supplementation on oxidative stress and antioxidant enzymes in older women. *Scand. J. Med. Sci. Sports* **2019**, *29*, 1101–1108. [CrossRef]

77. Sugihara Junior, P.; Ribeiro, A.S.; Nabuco, H.C.G.; Fernandes, R.R.; Tomeleri, C.M.; Cunha, P.M.; Venturini, D.; Barbosa, D.S.; Schoenfeld, B.J.; Cyrino, E.S. Effects of Whey Protein Supplementation Associated With Resistance Training on Muscular Strength, Hypertrophy, and Muscle Quality in Preconditioned Older Women. *Int. J. Sport Nutr. Exerc. Metab.* **2018**, *28*, 528–535. [CrossRef]

78. Mori, H.; Tokuda, Y. Effect of whey protein supplementation after resistance exercise on the muscle mass and physical function of healthy older women: A randomized controlled trial. *Geriatr. Gerontol. Int.* **2018**, *18*, 1398–1404. [CrossRef]

79. Englund, D.A.; Kirn, D.R.; Koochek, A.; Zhu, H.; Travison, T.G.; Reid, K.F.; von Berens, A.; Melin, M.; Cederholm, T.; Gustafsson, T.; et al. Nutritional Supplementation With Physical Activity Improves Muscle Composition in Mobility-Limited Older Adults, The VIVE2 Study: A Randomized, Double-Blind, Placebo-Controlled Trial. *J. Gerontol. Ser. Abiological Sci. Med Sci.* **2017**, *73*, 95–101. [CrossRef]

80. Fielding, R.A.; Travison, T.G.; Kirn, D.R.; Koochek, A.; Reid, K.F.; von Berens, A.; Zhu, H.; Folta, S.C.; Sacheck, J.M.; Nelson, M.E.; et al. Effect of Structured Physical Activity and Nutritional Supplementation on Physical Function in Mobility-Limited Older Adults: Results from the VIVE2 Randomized Trial. *J. Nutr. Health Aging* **2017**, *21*, 936–942. [CrossRef]

81. Chalé, A.; Cloutier, G.J.; Hau, C.; Phillips, E.M.; Dallal, G.E.; Fielding, R.A. Efficacy of whey protein supplementation on resistance exercise-induced changes in lean mass, muscle strength, and physical function in mobility-limited older adults. *J. Gerontol. Ser. Abiological Sci. Med Sci.* **2013**, *68*, 682–690. [CrossRef] [PubMed]

82. Kirk, B.; Mooney, K.; Amirabdollahian, F.; Khaiyat, O. Exercise and Dietary-Protein as a Countermeasure to Skeletal Muscle Weakness: Liverpool Hope University–Sarcopenia Aging Trial (LHU-SAT). *Front. Physiol.* **2019**, *10*. [CrossRef] [PubMed]

83. Kirk, B.; Mooney, K.; Cousins, R.; Angell, P.; Jackson, M.; Pugh, J.N.; Coyles, G.; Amirabdollahian, F.; Khaiyat, O. Effects of exercise and whey protein on muscle mass, fat mass, myoelectrical muscle fatigue and health-related quality of life in older adults: A secondary analysis of the Liverpool Hope University-Sarcopenia Ageing Trial (LHU-SAT). *Eur. J. Appl. Physiol.* **2020**, *120*, 493–503. [CrossRef] [PubMed]

84. Niccoli, S.; Kolobov, A.; Bon, T.; Rafilovich, S.; Munro, H.; Tanner, K.; Pearson, T.; Lees, S.J. Whey Protein Supplementation Improves Rehabilitation Outcomes in Hospitalized Geriatric Patients: A Double Blinded, Randomized Controlled Trial. *J. Nutr. Gerontol. Geriatr.* **2017**, *36*, 149–165. [CrossRef] [PubMed]

85. Gade, J.; Beck, A.M.; Bitz, C.; Christensen, B.; Klausen, T.W.; Vinther, A.; Astrup, A. Protein-enriched, milk-based supplement to counteract sarcopenia in acutely ill geriatric patients offered resistance exercise training during and after hospitalisation: Study protocol for a randomised, double-blind, multicentre trial. *BMJ Open* **2018**, *8*, e019210. [CrossRef] [PubMed]

86. Gade, J.; Beck, A.M.; Andersen, H.E.; Christensen, B.; Ronholt, F.; Klausen, T.W.; Vinther, A.; Astrup, A. Protein supplementation combined with low-intensity resistance training in geriatric medical patients during and after hospitalization: A randomized, double-blind, multicenter trial. *Br. J. Nutr.* **2019**, 1–34. [CrossRef]

87. Mancuso, M.; Orsucci, D.; Logerfo, A.; Rocchi, A.; Petrozzi, L.; Nesti, C.; Galetta, F.; Santoro, G.; Murri, L.; Siciliano, G. Oxidative stress biomarkers in mitochondrial myopathies, basally and after cysteine donor supplementation. *J. Neurol.* **2010**, *257*, 774–781. [CrossRef]

88. Chanet, A.; Verlaan, S.; Salles, J.; Giraudet, C.; Patrac, V.; Pidou, V.; Pouyet, C.; Hafnaoui, N.; Blot, A.; Cano, N.; et al. Supplementing Breakfast with a Vitamin D and Leucine–Enriched Whey Protein Medical Nutrition Drink Enhances Postprandial Muscle Protein Synthesis and Muscle Mass in Healthy Older Men. *J. Nutr.* **2017**, *147*, 2262–2271. [CrossRef]

89. Ascenzi, F.; Barberi, L.; Dobrowolny, G.; Villa Nova Bacurau, A.; Nicoletti, C.; Rizzuto, E.; Rosenthal, N.; Scicchitano, B.M.; Musarò, A. Effects of IGF-1 isoforms on muscle growth and sarcopenia. *Aging Cell* **2019**, *18*, e12954. [CrossRef]

90. Rodondi, A.; Ammann, P.; Ghilardi-Beuret, S.; Rizzoli, R. Zinc increases the effects of essential amino acids-whey protein supplements in frail elderly. *J. Nutr. Health Aging* **2009**, *13*, 491–497. [CrossRef]

91. Coker, R.H.; Miller, S.; Schutzler, S.; Deutz, N.; Wolfe, R.R. Whey protein and essential amino acids promote the reduction of adipose tissue and increased muscle protein synthesis during caloric restriction-induced weight loss in elderly, obese individuals. *Nutr. J.* **2012**, *11*, 105. [CrossRef] [PubMed]

92. Nabuco, H.C.G.; Tomeleri, C.M.; Fernandes, R.R.; Sugihara Junior, P.; Cavalcante, E.F.; Cunha, P.M.; Antunes, M.; Nunes, J.P.; Venturini, D.; Barbosa, D.S.; et al. Effect of whey protein supplementation combined with resistance training on body composition, muscular strength, functional capacity, and plasma-metabolism biomarkers in older women with sarcopenic obesity: A randomized, double-blind, placebo-controlled trial. *Clin. Nutr. ESPEN* **2019**, *32*, 88–95. [CrossRef] [PubMed]

93. Liberman, K.; Njemini, R.; Luiking, Y.; Forti, L.N.; Verlaan, S.; Bauer, J.M.; Memelink, R.; Brandt, K.; Donini, L.M.; Maggio, M.; et al. Thirteen weeks of supplementation of vitamin D and leucine-enriched whey protein nutritional supplement attenuates chronic low-grade inflammation in sarcopenic older adults: The PROVIDE study. *Aging Clin. Exp. Res.* **2019**, *31*, 845–854. [CrossRef] [PubMed]

94. Bo, Y.; Liu, C.; Ji, Z.; Yang, R.; An, Q.; Zhang, X.; You, J.; Duan, D.; Sun, Y.; Zhu, Y.; et al. A high whey protein, vitamin D and E supplement preserves muscle mass, strength, and quality of life in sarcopenic older adults: A double-blind randomized controlled trial. *Clin. Nutr.* **2019**, *38*, 159–164. [CrossRef] [PubMed]

95. Li, C.-W.; Yu, K.; Shyh-Chang, N.; Li, G.-X.; Jiang, L.-J.; Yu, S.-L.; Xu, L.-Y.; Liu, R.-J.; Guo, Z.-J.; Xie, H.-Y.; et al. Circulating factors associated with sarcopenia during ageing and after intensive lifestyle intervention. *J. Cachexiasarcopenia Muscle* **2019**, *10*, 586–600. [CrossRef]

96. Hill, T.R.; Verlaan, S.; Biesheuvel, E.; Eastell, R.; Bauer, J.M.; Bautmans, I.; Brandt, K.; Donini, L.M.; Maggio, M.; Mets, T.; et al. A Vitamin D, Calcium and Leucine-Enriched Whey Protein Nutritional Supplement Improves Measures of Bone Health in Sarcopenic Non-Malnourished Older Adults: The PROVIDE Study. *Calcif. Tissue Int.* **2019**, *105*, 383–391. [CrossRef]

97. Bauer, J.M.; Verlaan, S.; Bautmans, I.; Brandt, K.; Donini, L.M.; Maggio, M.; McMurdo, M.E.; Mets, T.; Seal, C.; Wijers, S.L.; et al. Effects of a vitamin D and leucine-enriched whey protein nutritional supplement on measures of sarcopenia in older adults, the PROVIDE study: A randomized, double-blind, placebo-controlled trial. *J. Am. Med. Dir. Assoc.* **2015**, *16*, 740–747. [CrossRef]

98. Il'yasova, D.; Scarbrough, P.; Spasojevic, I. Urinary biomarkers of oxidative status. *Clin. Chim. Acta* **2012**, *413*, 1446–1453. [CrossRef]

99. Cornish, S.M.; Candow, D.G.; Jantz, N.T.; Chilibeck, P.D.; Little, J.P.; Forbes, S.; Abeysekara, S.; Zello, G.A. Conjugated linoleic acid combined with creatine monohydrate and whey protein supplementation during strength training. *Int. J. Sport Nutr. Exerc. Metab.* **2009**, *19*, 79–96. [CrossRef]

100. Marrocco, I.; Altieri, F.; Peluso, I. Measurement and Clinical Significance of Biomarkers of Oxidative Stress in Humans. *Oxid. Med. Cell Longev.* **2017**, *2017*, 6501046. [CrossRef]

 *foods*

Review

Chicken Egg Proteins and Derived Peptides with Antioxidant Properties

Sara Benedé * and Elena Molina

Instituto de Investigación en Ciencias de la Alimentación (CIAL, CSIC-UAM), 28049 Madrid, Spain; e.molina@csic.es
* Correspondence: s.benede@csic.es

Received: 6 May 2020; Accepted: 1 June 2020; Published: 3 June 2020

Abstract: In addition to their high nutritional value, some chicken egg proteins and derivatives such as protein hydrolysates, peptides and amino acids show antioxidant properties which make them prominent candidates for the development of functional foods, drawing attention to both the food and biopharmaceutical industries. This review summarizes current knowledge on antioxidant activity of chicken egg proteins and their derived peptides. Some egg proteins such as ovalbumin, ovotransferrin and lysozyme from egg white or phosvitin from yolk have shown antioxidant properties, although derived peptides have higher bioactive potential. The main process for obtaining egg bioactive peptides is enzymatic hydrolysis of its proteins using enzymes and/or processing technologies such as heating, sonication or high-intensity-pulsed electric field. Different in vitro assays such as determination of reducing power, DPPH and ABTS radical-scavenging activity tests or oxygen radical absorbance capacity assay have been used to evaluate the diverse antioxidant mechanisms of proteins and peptides. Similarly, different cell lines and animal models including zebrafish, mice and rats have also been used. In summary, this review collects all the knowledge described so far regarding egg proteins and derived peptides with antioxidant functions.

Keywords: egg white; egg yolk; antioxidant peptides

1. Introduction

Eggs are not usually considered as antioxidant foods, however, many of their compounds such as vitamin E and A, selenium, phospholipids and carotenoids exhibit antioxidant properties [1]. In addition to their high nutritional value, chicken egg proteins and related ingredients (protein hydrolysates, peptides and amino acids) show several biological activities, including antioxidant activity, and therefore their use as functional and nutritional ingredients in food products has increased in recent years, drawing the attention of both the food and biopharmaceutical industries [2]. Moreover, natural antioxidants are considered safer for consumers than synthetic antioxidants and therefore there is a growing interest in them.

Egg white is mainly composed of protein (11%), being ovalbumin the most abundant (54%) followed by ovotransferrin (12%), ovomucoid (11%), lysozyme (3.5%), and ovomucin (3.5%). Besides, other minor proteins such as avidin, cystatin, ovomacroglobulin, ovoflavoprotein, ovoglycoprotein and ovoinhibitor have also been identified [3]. The main components of the yolk are lipids (31–35%) although it also has 15–17% of proteins including lipovitellins (36%), livetins (38%), phosvitin (8%), and low-density lipoproteins (17%) [4]. Egg yolk is covered with the vitelline membrane which separates it from the egg white and it is also a good source of proteins, composed mostly of protein fibers [5].

Some of these proteins have antioxidant properties by themselves but it has been demonstrated that peptides derived from them, usual fragments of 2–20 amino acid residues, have the higher bioactive potential [6]. The hypothesis that arises to explain this fact is that small peptides have increased accessibility of the functional side chain (R-group) to the reactive species and the electron-dense peptide bonds and therefore they can exert their antioxidant function more easily [7]. However, not only peptide length is associated with the activity of antioxidant peptides and amino acid composition seems to have an important role. The sulfur-containing amino acids such as cysteine and methionine are prone to oxidation due to their S-groups that form stable oxidation products by reacting with reactive species. Acidic amino acids such as glutamine and asparagine and the hydrophobic proline, alanine, cysteine, valine, methionine, isoleucine, leucine, tyrosine, phenylalanine and tryptophan amino acids have also a strong positive effect on antioxidant activity [7]. Functional peptides are not active within the sequence of the parent protein molecule and can be released by in vivo or in vitro processes [8]. The main procedure for obtaining bioactive peptides from food products is enzymatic hydrolysis of proteins orchestrated with the use of various enzymes of microbial, plant or animal origin [9], although chemical hydrolysis [10] or processing technologies such as heat or high-intensity pulsed electric field treatments can also be applied [11,12].

Antioxidant peptides from eggs can inhibit oxidative stress, which plays an important role in human health, and in food systems, and increase the quality and the shelf life of products. Moreover, antioxidant peptides prevent oxidative damages through multiple pathways such as free radical scavenging, chelating pro-oxidative transition metal ions, inactivation of reactive oxygen species and reducing hydroperoxides [9,13,14] and, therefore, there is not a single antioxidant test model to evaluate their activity. In practice, several in vitro assays have been developed to quantify antioxidant activities and can be classified into two types; hydrogen atom or electron transfer reaction-based assays [15]. Hydrogen atom transfer reaction-based assays quantify the hydrogen atom donating ability of the antioxidant compound resulting in a kinetic curve, while the electron transfer reaction-based assays measure the reducing capacity of the antioxidant compound, resulting in a color shift that can be measured by the change in absorbance [16]. The 1,1-diphenyl-2-picrylhydrazyl (DPPH) radical scavenging activity and reducing power assays are the most used in studies on the antioxidant capacity of egg derivatives [17]. Among the hydrogen atom transfer-based method most commonly used methods to measure the antioxidant capacity of egg-derived peptides is oxygen radical absorbance capacity (ORAC) which consist in a fluorescent compound, such as fluorescein, that is damaged by free radicals and subsequently loses its fluorescence, but when antioxidants are present the loss of fluorescence is inhibited [18]. The 2, 2-azinobis (3-ethyl benzothiazoline-6-sulfonic acid) diamonium salt (ABTS) radical scavenging method and lipid peroxidation inhibition assay are also included in this category, among others [17,19]. Despite the numerous *in vitro* studies performed to evaluate the antioxidant capacity of egg-derived hydrolysates or peptides, their commercial application is delayed due to the lack of scalable production processes, the few digestibility and bioavailability studies as well as animal studies available and the absence of clinical trials that probe their potential health benefits. In this work, current knowledge on antioxidant activity of chicken egg proteins, the main strategies to obtain antioxidant peptides from them and the identified peptides are summarized.

2. Antioxidant Activity of Chicken Egg Proteins

2.1. Ovalbumin

Ovalbumin is the most abundant protein in egg white. It has a molecular mass of 45 kDa and is compiled of 385 amino acids. It contains one disulfide bond and four free sulphydryl groups, which can play a role in redox regulation, acting as metal chelators [20]. Several studies have described an increased antioxidant capacity of ovalbumin after covalent binding. Glycosylation of ovalbumin with glucose under heat moisture treatment [21] or microwave heating [22] increased its antioxidant activity measured by determination of DPPH radical-scavenging activity and Trolox equivalent antioxidant capacity assay. Glycation with mannose by ultrasound technology [23], maltose by heat treatment [21] or covalent binding with dextran or galactomannan by a controlled Maillard reaction [24] also exhibited higher antioxidant activity than native ovalbumin. In addition, the raising of antioxidant activity has also been identified when ovalbumin was combined with the polyphenol rutin [25] or the mineral selenite [26] which was attributed to the formation of a molten globule conformation of the protein, increasing its surface hydrophobicity and therefore its solubility.

2.2. Ovotransferrin

Ovotransferrin is composed of 686 amino acids and has a molecular mass of 77.90 kDa. It is folded into two globular lobes with an iron-binding site and interconnected by an alpha-helix of nine amino acidic residues. Fifteen disulfide bridges stabilize the structure [27]. In addition to antimicrobial activity, antioxidant properties have also been attributed to ovotransferrin [10]. Binding of ovotransferrin with metals such as iron, magnesium and copper [28], conjugation with small molecules such as catechin [29] or autoclaved treatment improved its antioxidant activity [30].

2.3. Lysozyme

Lysozyme contains 129 amino acids, presents a molecular mass of 14.3 kDa and four disulfide bridges. It inhibits reactive oxygen species generation [31] and like other egg white proteins, its antioxidant properties increased after conjugation with other compounds such as polysaccharides [32]. Conjugation of lysozyme with guar gum, a hydrophilic polysaccharide extracted from the seeds of *Cyamopsis tetragonolobus* increased its antioxidant properties from 2% to 35% of inhibition of DPPH [33]. The alkaline pH used for the preparation of the conjugate opened the globular structure of the protein, causing electron-donating amino acid residues to get more exposed and therefore increasing lysozyme reducing power. The same effect was observed when xanthan gum, an anionic extracellular polysaccharide secreted by the microorganism *Xanthomonas campestris* was used for the conjugation [34].

2.4. Cystatin

Cystatin is a small protein of approximately 13 kDa molecular weight which contains two disulfide bonds. It is been shown that it modulates the synthesis and release of nitric oxide production in murine macrophages and thereby plays a role in cellular antioxidant pathways [35]. Optimum levels of nitric oxide are essential for the regulation of specific cellular antioxidant pathways. In addition, cystatin could protect brain neurons from oxidative damage [36].

2.5. Phosvitin

Phosvitin is the major protein component in egg yolk. It has a sequence of 216 amino acid residues that contains 123 serine residues of which most are phosphorylated. Phosvitin has high chelating power for cations which has led to several studies that prove its great antioxidant properties [37,38]. Moreover, free aromatic amino acids such as tryptophan and tyrosine in egg yolk were found to be largely responsible for the antioxidant properties of egg yolk [39].

3. Production of Antioxidant Peptides from Chicken Egg Proteins

Bioactive peptides from eggs have been mainly produced from egg white proteins, although egg yolk has also recently been used as a new source of functional peptides, as well as other egg components such as eggshell or vitelline membrane. The whole egg white and yolk or a single protein can be used as starting material to produce bioactive peptides. The main procedure used to obtain peptides has been hydrolysis with food-grade proteolytic enzymes from animal, plant or bacterial origin [9]. Commercial enzymes are preferred over naturally occurring ones because their specific characteristics such as optimal pH, temperature or cleavage site are well defined [40]. Normally, one or several proteases are used to obtain proteins hydrolysate [41] being pepsin, trypsin, alcalase and papain some of the most popular enzymes used [12,42–44]. Alternatively, non-commercial enzymes have also applied to produce functional egg peptides to reduce the cost of hydrolysis [41,45].

Modification of egg proteins previously to the hydrolysis process results in protein conjugates that lead to increase radical scavenging properties [46] and therefore several physical methods such as high-intensity pulsed electric field [12], sonication [11], heat [12] and high pressure [47] treatments have been applied to increase enzymatic digestibility. The advantage of hydrolysis is that it is easy to scale up but there are numerous variables involved in the production and purification of antioxidant peptides and optimization is not taken into account in most of the published literature.

Once the hydrolysate has been obtained, the use of techniques such as ultrafiltration or liquid chromatography allows it to be separated into different fractions that can be evaluated to determine their antioxidant potential. The techniques most widely used for the measurement of antioxidant capacity are in vitro colorimetric assays in which samples compete with substrates for the radicals and inhibits or restrict the substrate oxidation [48]. After the identification of hydrolysate fractions with antioxidant properties, the generated peptides can be identified by mass spectrometry and chemically synthesized for validation.

3.1. Hydrolysis of Egg White

Pepsin is one of the most used enzymes to obtain egg white hydrolysates. Dávalos et al. described that 3-h proteolysis of crude egg white at pH 2 and 37 °C with an enzyme to substrate ratio of 1/100 increased the radical scavenging activity by approximately threefold compared to untreated crude egg white, being the fraction lower than 3 kDa the most active, probably because of the higher accessibility of small peptides to the redox reaction system [49]. They identified four peptides from ovalbumin with higher radical scavenging activity than that of Trolox and with lipid peroxidation inhibition ability (Table 1). Their increased antioxidant activity was imputed to the presence of tyrosine at the N terminus as the presence of a hydroxyl group in the tyrosine aromatic structure allows it to break the antioxidant chain by a hydrogen atom transfer mechanism [50].

Table 1. Sequences of antioxidant peptides obtained by hydrolysis of chicken egg white proteins.

Peptide Sequence	Protein of Origin	Starting Material	Enzymes	Antioxidant Assay	References
RVPSLM TPSPR DLQGK AGLAPY RVPSL	OVT	EW	Alcalase	DPPH	[51]
IRW LKP	OVT	OVT	Thermolysin Pepsin	ORAC	[52]
WNIP GWNI	OVT	OVT	Thermolysin	ORAC	[11]
YAEERYPIL SALAM YQIGL YRGGLEPINF	OVA	EW	Pepsin	DPPH, LP	[49]
DHPFLF HAEIN QIGLF	OVA	EW	Alcalase	DPPH	[51]
AEERYP AEERYP DEDTQAMP	OVA	EW	Protease P	ORAC	[1]
NTDGSTDYGILQINSR	LZ	LZ	Papain Trypsin	DPPH	[53]
RGY WIR VAW	LZ	LZ	Pepsin Trypsin α-chymotrypsin	RP, LP	[54]

Table 1. *Cont.*

Peptide Sequence	Protein of Origin	Starting Material	Enzymes	Antioxidant Assay	References
VAWRNRCKGTD IRGCRL WIRGCRL AWIRGCRL WRNRCKGTD	LZ	LZ	Pepsin	ORAC, TBARS in Zebrafish larvae	[55]
WNWAD	OM	EW	Pepsin	ORAC, HEK-293 cells	[56]
PVDENDEG	CY	EW	Protease P	ORAC	[1]
HANENIF VKELY TNGIIR	EW	EW	Alcalase	DPPH	[51]
HTKE FFGFN MPDAH DHTKE	EW	EW	Alcalase	RP, DPPH, ABTS, ORAC	[57]
VYLPR EVYLPR VEVYLPR VVEVYLPR	EW	EW	Alcalase	ORAC, ABTS, H2O2-induced oxidative damage on HEK-293	[44]
YLGAK GGLEPINFN	EW	EW	Papain		[58]

OVT: ovotransferrin; EW: egg white; OVA: Ovalbumin; LZ: Lysozyme; OM: Ovomucoid; CY: Cystatin; DPPH: 2,2-difenil-1-picrylhydrazyl radical- scavenging activity assay; ORAC: Oxygen radical absorbance assay; RP: Reducing power assay; ABTS: 2,2'-azino-bis(3-ethylbenzothiazoline-6-sulfonic acid) radical-scavenging activity assay; TBARS: Thiobarbituric acid reactive substances; LP: Lipid peroxidation inhibition assay.

In addition, the presence of methionine in the peptide structure could also be conducive to the increase of the antioxidant activity because cyclic oxidation and reduction of methionine is an important antioxidant mechanism [59]. The same 3-h protocol was used to produce egg white hydrolysate with pepsin and examined the effects of its long-term consumption on spontaneously hypertensive rats, which make not only an accepted animal model for human hypertension, but also for oxidative stress because lipid peroxidation directly damages cell membranes and increases the levels of vasoconstrictor hormones such as angiotensin and endothelin which directly induce hypertension [60]. Pepsin hydrolysate was effective in increasing the radical-scavenging capacity of the plasma and decreasing the malondialdehyde concentration, a biomarker of oxidative damage, in the aorta [61]. In addition, pepsin egg white hydrolysate has also been administrated to an experimental model of obesity using Zucker fatty rats. Obesity is associated with abnormal production of proinflammatory mediators by the fat tissue and the pepsin hydrolysate also reduced levels of plasma malondialdehyde [62].

In accordance with Dávalos et al., other study probed that DPPH, hydroxyl and superoxide anion radical-scavenging activities of the hydrolysates obtained with pepsin depend on the molecular weight of the generated peptides, being the fraction with 2–5 kDa peptides the one with stronger antioxidant activity [63], although it is difficult a precise comparison because of the different hydrolysis conditions used (Table 2). A wide screening of the conditions required to obtain the pepsin hydrolysates with the strongest antioxidant capacity was done by Lin et al. which determined that 4.56% of egg white as starting material with an enzyme to substrate ratio of 1.58% at pH 1.99 and 37 °C for 1 h were the optimal conditions, although they do not specify the activity of the pepsin used and a pre-treatment of 10 min at 90 °C was applied to denature egg white proteins [12].

Table 2. Hydrolysis conditions applied to obtain antioxidant egg white hydrolysates using pepsin at pH 2.

Starting Material	Pre-Treatment	Pepsin Activity	Enzyme to Substrate Ratio	T (°C)	Time	Stop Conditions	References
Dissolved EW (100 mg/mL)	-	10,000 U/mg	1/100 (*w/w*)	37	3 h	pH 7	[49,61]
Pasteurized EW	-	3000 U/mg	2:100 (*w/w*)	38	8 h	pH 7	[62]
Freeze-dried EW	90 °C, 10 min	9000 U/g	30 g/L	37	5 h	85 °C, 30 min	[63]
EW powder 5.56%	90 °C, 10 min	NS	1.58% (*w/w*)	37	1 h	90 °C, 10 min	[12]
Liquid EW in water (1/4)	95 °C, 10 min	NS	0.4% (*w/v*)	37	1 h	Heat (NS)	[43]
Pasteurized EW	-	3000 U/mg	2:100 (*w/w*)	38	48 h	95 °C, 15 min	[64]

T: temperature; EW: egg white; NS: do not specify.

Other studies have also probed the antioxidant capacity of hydrolyzed egg white with pepsin using other methods such as Ferric Reducing Antioxidant Power (FRAP) assay [43] or measurement of oxidative stress inhibitory activity in cell lines such as in the study Garcés-Rimón et al., where they observed a dose-dependent inhibition of reactive oxygen species production in a macrophage cell line after the treatment with an egg white hydrolysate obtained with pepsin [64].

Trypsin is another common enzyme to obtain bioactive peptides with antioxidant functions [43,58,65]. The optimal enzymatic parameters that have been described are 4.93% of egg white as the starting material with an enzyme to substrate ratio of 1.61% at pH 9.05 and 37 °C for 1 h [12]. Similarly, the optimal conditions of egg white hydrolysis with alcalase have been described as 5% of egg white protein powder as the starting material with an enzyme-to-substrate ratio of 3% at pH 11 and 50 °C for 3 h, being the hydrolysate fraction containing peptides with a molecular mass lower than 1 kDa the one with the highest antioxidant activity [66]. Similar results using the same hydrolysis protocol were reported later proving that antioxidant activity of peptides is intimately connected to their molecular weight [57]. Furthermore, other studies have also reported increased antioxidant activity of egg white hydrolysate with alcalase, although using different hydrolysis conditions [43,67–69].

Antioxidant peptides obtain from hydrolysis of egg white with alcalase have been identified by several authors. Yu et al. identified 11 peptides (Table 1) with DPPH radical-scavenging activity [51], and Liu et al. identified 4 peptides (Table 1) and in addition, observed that antioxidant capacity was restricted by the amino acid composition of peptides indicating that Leucine, aspartic acid, serine, glutamic acid and lysine could play an important role [57]. Moreover, a recent study that identifies four more peptides (Table 1), reported that valine at the N-terminus is useful to increases the antioxidant activity of peptides [44]. Furthermore, this study used the peptide VYLPR to investigate the antioxidant mechanism on HEK-293 cells. They observe that VYLPR could inhibit lipid peroxidation, contribute to cell membrane integrity, prevent intracellular lactate dehydrogenase activity, which is increased under the condition of oxidative stress, reduce the oxidative biomarker malondialdehyde, and improve the activity well known antioxidant enzymes such as superoxide dismutase and glutathione peroxidase [44].

Hydrolysis of egg white by papain can also produce hydrolysates with the ability to quench the superoxide anion and hydroxyl radicals, prevent lipid peroxidation and show reducing power [58]. Two peptides, YLGAK and GGLEPINFN (Table 1) showed strong antioxidant activity in DPPH radical scavenging and lipid peroxidation inhibition tests [58]. Other commercial enzymes have also been used to obtain egg white hydrolysates with antioxidant properties. Neutrase, protamex, collupulin, ficin, flavourzyme, protease M and protease P have been successfully used to obtain egg white hydrolysates with high radical scavenging activity [1,65,67]. In fact, four peptides (Table 1), obtained after hydrolysis of egg white proteins with protease P and with high oxygen radical absorbance capacity have been identified [1].

In addition to commercial enzymes, there is growing attention in the finding of microbial proteases generated by fermentation procedures. In this line, some recent studies have purified fungal proteases and subsequently used to hydrolyze egg white, resulting in hydrolysates with high antioxidant activity. Garcés-Rimón et al. used flavourzyme 1000 L and peptidase 433 P to hydrolyze commercial pasteurized egg white and obtained hydrolysates with a high oxygen radical absorbance capacity. Moreover, these results were confirmed in a macrophage cell line where they showed a dose-dependent inhibition of reactive oxygen species generation [64]. A new enzyme produced by *Aspergillus avenaceus* URM 6706 has been used in the hydrolysis of egg white at pH 10.0 and 50 °C and a positive correlation between the in vitro antioxidant activity and the degree of hydrolysis has been observed [41]. Similar results were obtained when fungal proteases obtained from *Eupenicillium javanicum* and *Myceliophthora thermophile* were used [45].

A combination of different enzymes can also be used to produce egg white hydrolysates. Different combinations of pepsin, chymotrypsin and Alcalase 2.4 L to performed double enzyme hydrolysis of egg white could produce highly antioxidative peptides [43]. The hydrolysate with higher antioxidant activity was obtained with the combined use of pepsin and chymotrypsin because, due to their cutting sites, a hydrolysate with more aromatic aminoacid residues was obtained and these aromatic residues can quench the free radicals by electron transfer [43].

Due to the fact that smaller antioxidant peptides could exert better biological effects [70], technological treatments have been applied before or after enzymatic hydrolysis of egg white with the purpose to increase the decomposition of proteins. The most commonly used treatment is heat to denature proteins prior to hydrolysis and facilitate access to enzymes [12,43,51]. Microwave, high-pressure and ultrasound pre-treatments of egg white have also been applied to obtain hydrolysates with higher antioxidant activity after pepsin, trypsin and alcalase hydrolysis, respectively [22,71,72]. Other treatments such as pulsed electric fields can also be applied after the hydrolysis process. Lin et al. used this technology to treat the fraction containing peptides with a molecular mass lower than 1 kDa of an egg white hydrolysate obtained with alcalase [66]. As a result, the antioxidant activity of the treated peptide fraction was increased due to the higher number of small peptides and the exposure of histidine, proline, cysteine, tyrosine, tryptophan, phenylalanine, and methionine residues.

Proteolytic degradation of egg white during the digestion process can also promote the formation of antioxidant peptides as it has been proven in several studies using simulated gastrointestinal digestions assays [73,74]. Measurement of antioxidant capacity of cooked eggs after simulated gastrointestinal digestion indicated that although the cooking of eggs reduced their antioxidant activity, the generated peptides showed higher antioxidant activity, being three peptides derived from ovalbumin (DSTRTQ, DVYSF and ESKPV) identified with antioxidant activity in a smooth muscle cell line [74]. This study used a dynamic system to mimic conditions in the gastrointestinal tract, comprising four compartments that represent the stomach, duodenum, jejunum and ileum. Stability during digestion is important to ensure the bioavailability of bioactive peptides and obtain the desired activity when tested in vivo [75]. Similarly, Jahandideh et al., also observed a reduction in tissue oxidative stress in spontaneously hypertensive rats after the administration of fried egg white previously digested in a simulated gastro-intestinal digestion system using pepsin and pancreatin [76]. However, it should be noticed that more research is needed to determine the effect of digestion on antioxidant peptides release from egg proteins because differences have been observed when in vivo research is performed. As an example, fried whole egg previously digested with commercial pepsin and pancreatin reduced tissue oxidative stress in spontaneously hypertensive rats, but the same effect was not observed when non-hydrolyzed fried whole egg was administrated to the rat despite it also underwent digestion in the digestive tract of the animals [76]. These opposite results could be attributed to differences in the origin of enzymes, time of hydrolysis, pH or temperature of hydrolysis, among others, and should be taken into account if commercially production of these peptides is desired to avoid scaling-up issues.

3.2. Hydrolysis of Egg Yolk

Egg yolk is also a rich generator of antioxidants due to the presence of free aromatic amino acids, being tryptophan and tyrosine two of the main contributors to the antioxidant properties of egg yolk [39]. In addition, as with egg white, antioxidant peptides can be obtained after enzymatic hydrolysis of egg yolk. Several commonly used enzymes such as pepsin, trypsin and chymotrypsin have been effective to produce egg yolk hydrolysates with antioxidant activity [77–79] and with the capacity to protect DNA against oxidative damage induced by peroxide [80]. The hydrolysis of defatted egg yolk with pepsin followed by pacreatin could be used to reduce stress oxidative in spontaneously hypertensive rats [76]. Other studies have identified the sequence of several egg yolk peptides derived from proteins hydrolyzsated with pepsin. Yousr and Howell, identified three peptide sequences (WYGPD, KLSDW and KGLWE) with the capacity to inhibit the peroxides and thiobarbituric acid reactive molecules in an oxidizing linoleic acid model system [81], being the superoxide anion and hydroxyl radicals scavenging and ferrous chelation the antioxidant mechanisms involved, although hydrophobic amino acids such as tyrosine and tryptophan in identified sequences could also influence [81]. Hydrolysis of egg yolk with pepsin gave rise to four peptides from Apolipoprotein B (YINQMPQKSRE; YINQMPQKSREA), Vitellogenin-2 (VTGRFAGHPAAQ) and Apovitellenin-1 (YIEAVNKVSPRAGQF) with in vitro antioxidant activity [82]. A combination of high hydrostatic pressure treatment and enzymatic hydrolysis with alcalase, elastase, savinase, thermolysin or trypsin has also been applied successfully to produce peptides derived from phosvitin with higher antioxidant activity than the native protein [83].

Phosvitin phosphopeptides obtain after hydrolysis of egg yolk with trypsin followed by an alkaline dephosphorylation of phosvitin were effective to reduce oxidative stress in an in vitro system of human intestinal epithelial cells [84,85]. In addition, IL-8 released after treatment of cells with H_2O_2 was reduced [86]. The authors attributed the strong antioxidant activity of phosvitin phosphopeptides to their rich amino acid composition in histidine, methionine and tyrosine rather than the presence of phosphorylserine ligands [87]. Phosvitin phosphopeptides have also been obtained using alcalase, being able to reduce in vitro oxidative stress by up-regulating glutathione synthesis and antioxidant enzyme activities [88]. Moreover, this activity is maintained after gastrointestinal digestion and can promote the antioxidant capacity of enzymes such as catalase and glutathione S-transferase, and reduced protein and lipid oxidation in the intestine of a porcine model of oxidative stress [89]. Alcalase was also used by Park et al., to hydrolyze egg yolk protein, obtained as a secondary product after purification of lecithin, and identify two antioxidant peptides (LMSYMWSTSM and LELHKLRSSHWFSRR). The activity of these peptides was attributed to the presence of leucine at their N-terminal positions [90].

Hydrolysis of egg yolk protein with a serine proteinase from Asian pumpkin pulp for 4 h allowed the production of a hydrolyzate with DPPH free radical scavenging capacity [91], as well as the identification of four peptides (RASDPLLSV, RNDDLNYIQ, LAPSLPGKPKPD and AGTTCLFTPLALPYDYSH) with in vitro antioxidant activity analyzed by DPPH scavenging activity, ferric reducing ability and ferrous ion-chelating activity [9]. The use of other enzymes such as serine protease from yeast has also allowed the identification of QSLVSVPGMS peptide that exhibits a high in vitro DPPH free radical scavenging activity [92].

Other enzymes from the microbial origin such as neutrase, thermolysin and pronase, among others, have also been used to produce hydrolysates with in vitro DPPH scavenging and chelating iron activity [77,93,94]. Sakanaka et al. reported that egg yolk protein hydrolysates, obtained with the action of proteinase from *Bacillus* sp., display antioxidant activities in a linoleic acid oxidation system [77]. They proved its antioxidant capacity on cookies containing linoleic acid and, therefore, was proposed as a natural antioxidant for avoiding the oxidation of polyunsaturated fatty acids in food products. In addition, egg yolk peptides have been also useful to inhibit lipid oxidation in other food matrices such as beef or tuna muscle homogenates [93]. In fact, as a natural protein derived from animal products, phosvitin and its derived peptides have been proposed to be used as antioxidant in meat products [95].

3.3. Hydrolysis of Other Egg Components

Vitelline membrane, a multilayered structure that surrounds the egg yolk separating the yolk and white, is composed of about 87% protein. Its DPPH scavenging, superoxide radical scavenging and iron-chelating activities have been proven in in vitro studies. Moreover, the antioxidant capacity of the vitelline membrane was improved after enzymatic hydrolysis with alcalase, flavourzyme or trypsin [13]. Hydrolysates from eggshell membrane proteins have been obtained using alcalase and protease S [96] with the potential to suppress lipid and protein oxidation against oxidative stress damage induced by H_2O_2 in Caco-2 cells. The mechanisms involved in this effect were the elevation of antioxidant enzyme activities and cellular levels of GSH, a cellular endogenous antioxidant, via up-regulation of its mRNA expression. In addition, these hydrolysates increased γ-GCS activity, which catalyzes GSH synthesis from glutamate, cysteine and glycine [97].

4. Peptides from Individual Egg White Proteins

Foods contain many naturally occurring compounds [98] that can interact with the proteins in the matrix, affecting the type of peptides generated upon hydrolysis and should be considered during the production process of antioxidant peptides [75]. Despite this fact, functional peptides have been obtained from individual egg white proteins in many studies.

4.1. Ovotransferrin

After digestion of ovotransferrin by thermolysin and pepsin, the resultant hydrolysate showed higher oxygen radical absorbance value than the native protein. The peptide IRW, derived from hydrolysis, exhibited a high oxygen radical-scavenging effect, which might be attributed to the presence of tryptophan [52]. Moreover, two tetrapeptides (WNIP and GWNI) with high antioxidant activity have been identified within the digests of ovotransferrin with thermolysin. The motif of WNI seemed to be responsible for the high antioxidant capacity because amino acid residues coupled to either the N or C terminus of both peptides reduced their antioxidant capacity [11]. In addition, the peptide GWNI has shown the capacity to reduce reactive oxygen species generation when it was tested in endothelial cells [99]. In contrast, other antioxidant peptides previously identified from ovotransferrin using the oxygen radical absorbance capacity did not exhibit any antioxidant activity in cells, showing the deficiencies of cell-free in vitro methods for antioxidant studies and highlighting the need to use more biological systems such as culture cells to evaluate antioxidant peptides [99]. Hydrolyzates derived from ovotransferrin obtained with HCl at pH 2.5 or other different enzymes such as protamex, alkalase, trypsin, neutrase, flavorzyme, maxazyme, collupulin, protex, promod 278, and alpha-chymotrypsin showing higher superoxide anion scavenging activity and oxygen radical absorbance capacity than intact protein and demonstrating preventive effects against the oxidative stress-induced DNA damage in human leukocytes [10].

4.2. Lysozyme

Lysozyme hydrolyzed with alcalase showed similar oxygen radical absorbance activity than the synthetic antioxidants hidroxibutilanisol and butilhidroxitolueno in foods [100]. Lysozyme hydrolyzed with papain, trypsin or a combination of the two enzymes showed high DPPH scavenging activity and the peptide NTDGSTDYGILQINSR, resulting from the double hydrolysis, was identified as responsible of the hydrolysate antioxidant capacity [53]. Gastrointestinal digestion of lysozyme may also induce the generation of antioxidant peptides. Lysozyme subjected to simulated physiological digestion using pepsin, trypsin and alpha-chymotrypsin was found to have potent antioxidant capacity determined by DPPH radical scavenging and reducing power assays. In addition, three peptides with strong antioxidant capacity were identified as products of digestion [54]. Hydrolysis of lysozyme with pepsin can give rise to antioxidant peptides. Carrillo et al. identified five positively charged peptides (Table 1) with high oxygen radical absorbance capacity in vitro. They were one step further and confirmed the results in a Zebrafish larvae model testing oxidative stress by measuring the inhibition of lipid peroxidation [55]. For this, they incubated larvae with lysozyme peptides and initiated lipid peroxidation by adding H_2O_2. After incubation during 8 h at 28 °C, the Zebrafish larvae were homogenized and measured in a spectrophotometer. The drop of absorbance indicated an elevation of antioxidant activity.

4.3. Ovalbumin

Ovalbumin hydrolysate obtained by hydrolysis with pepsin was also tested in vivo using aged mice. The hydrolysate significantly decreased malondialdehyde content in the serum and liver of mice, proving its antioxidative activity [101]. Moreover, hydrolyzing ovalbumin with different combinations of enzymes such as pepsin and papain, pepsin and alcalase, alcalase and trypsin, and alpha-chymotrypsin were also effective in producing peptides with antioxidant capability as well as strong iron and copper-binding capacity [102].

4.4. Other Egg Proteins

Several enzymes alone such as pepsin and alcalase or in combination such as alcalase and trypsin or alcalase and trypsin have also been used to obtain ovomucoid hydrolysates, all of them showing strong antioxidant activity analyzed by the thiobarbituric acid reactive substances method [103]. Ovomucin hydrolysates obtained by the action of trypsin, papain or alcalase have also antioxidant activity. In addition, heating ovomucin under alkaline conditions gave rise peptides with strong iron-binding and antioxidant activities [104]. In addition, antioxidant peptide derived from the ultrafiltrate of ovomucin hydrolysate inhibit H_2O_2-induced oxidative stress in the human embryonic kidney [105]. Furthermore, oligophosphopeptides from phosvitin obtained after tryptic hydrolysis exhibited high antioxidant capacity in DPPH free-radical-scavenging tests [87] and in Caco-2 cells [86], showing a preventive effect against oxidation-induced DNA damage in vivo and preventing iron-mediated oxidative stress-related diseases, such as colorectal cancer [80].

5. Conclusions

Oxidative stress plays an important role in human health, and in food systems can increase the quality and the shelf life of products. Antioxidant compounds prevent oxidative and among them, natural antioxidants are considered safer for consumers than synthetic antioxidants, showing a growing interest in the last years. In addition to their high nutritional value, some chicken egg proteins and related ingredients (protein hydrolysates, peptides and amino acids) show antioxidant activity. There is extensive research on the evaluation of the antioxidant capacity of egg-derived hydrolysates and derived peptides. Many egg proteins such as ovalbumin, ovotransferrin, lysozyme, and cystatin from egg white or phosvitin from egg yolk are reported to have antioxidant properties by themselves. However bioactive peptides with increased antioxidant activity than proteins can be obtained after hydrolysis of egg proteins with proteolytic enzymes from animal, plant or bacterial origin as well as by chemical hydrolysis or gastrointestinal digestion. The whole egg white and yolk or a single protein are usually used as starting material to produce bioactive peptides and modification of egg proteins previously to the hydrolysis process using high-intensity pulsed electric field, sonication, heat and high-pressure treatments are used to increase radical scavenging properties. The sequences of several antioxidant peptides derived from the egg have been identified and their properties have been evaluated in chemical in vitro assays, cell cultures and animal models. However, a small number of pure peptides have been investigated at a cellular level and even less in in vivo systems, and moreover, there is a lack of digestibility and bioavailability studies as well as clinical trials. Therefore, further research is needed to recommend antioxidant egg-derived peptides for preventive and therapeutic treatments of both healthy subjects and patients with diseases related to stress oxidative. In addition, studies on the safety and quality of foods containing antioxidant peptides are also needed.

Author Contributions: Conceptualization, S.B. and E.M.; writing—original draft preparation, S.B.; writing—review and editing, S.B. and E.M.; supervision, E.M. All authors have read and agreed to the published version of the manuscript.

Acknowledgments: S.B. acknowledges the financial support of Ministerio de Ciencia, Innovación y Universidades from Spain through a Juan de la Cierva Incorporación contract.

Conflicts of Interest: The authors declare no conflicts of interest.

References

1. Nimalaratne, C.; Bandara, N.; Wu, J. Purification and characterization of antioxidant peptides from enzymatically hydrolyzed chicken egg white. *Food Chem.* **2015**, *188*, 467–472. [CrossRef] [PubMed]
2. Miranda, J.M.; Anton, X.; Redondo-Valbuena, C.; Roca-Saavedra, P.; Rodriguez, J.A.; Lamas, A.; Franco, C.M.; Cepeda, A. Egg and Egg-Derived Foods: Effects on Human Health and Use as Functional Foods. *Nutrients* **2015**, *7*, 706–729. [CrossRef] [PubMed]

3. Kovacs-Nolan, J.; Phillips, M.; Mine, Y. Advances in the Value of Eggs and Egg Components for Human Health. *J. Agric. Food Chem.* **2005**, *53*, 8421–8431. [CrossRef] [PubMed]

4. Abeyrathne, E.D.N.S.; Lee, H.Y.; Ahn, D.U. Egg white proteins and their potential use in food processing or as nutraceutical and pharmaceutical agents—A review. *Poult. Sci.* **2013**, *92*, 3292–3299. [CrossRef] [PubMed]

5. Mann, K. Proteomic analysis of the chicken egg vitelline membrane. *Proteomics* **2008**, *8*, 2322–2332. [CrossRef]

6. Martinez-Villaluenga, C.; Penas, E.; Frias, J. Bioactive peptides in fermented foods: Production and evidence for health effects. In *Fermented Foods in Health and Disease Prevention*; Frias, J., MartinezVillaluenga, C., Penas, E., Eds.; Academic Press Ltd-Elsevier Science Ltd: London, UK, 2017; pp. 23–27. [CrossRef]

7. Udenigwe, C.C.; E Aluko, R. Chemometric Analysis of the Amino Acid Requirements of Antioxidant Food Protein Hydrolysates. *Int. J. Mol. Sci.* **2011**, *12*, 3148–3161. [CrossRef]

8. Agyei, D.; Danquah, M.K. Industrial-scale manufacturing of pharmaceutical-grade bioactive peptides. *Biotechnol. Adv.* **2011**, *29*, 272–277. [CrossRef]

9. Zambrowicz, A.; Eckert, E.; Pokora, M.; Bobak, L.; Szołtysik, M.; Trziszka, T.; Chrzanowska, J.; Dąbrowska, A. Antioxidant and antidiabetic activities of peptides isolated from a hydrolysate of an egg-yolk protein by-product prepared with a proteinase from Asian pumpkin (Cucurbita ficifolia). *RSC Adv.* **2015**, *5*, 10460–10467. [CrossRef]

10. Kim, J.; Moon, S.H.; Ahn, D.U.; Paik, H.-D.; Park, E. Antioxidant effects of ovotransferrin and its hydrolysates. *Poult. Sci.* **2012**, *91*, 2747–2754. [CrossRef]

11. Shen, S.; Chahal, B.; Majumder, K.; You, S.-J.; Wu, J. Identification of Novel Antioxidative Peptides Derived from a Thermolytic Hydrolysate of Ovotransferrin by LC-MS/MS. *J. Agric. Food Chem.* **2010**, *58*, 7664–7672. [CrossRef]

12. Lin, S.; Jin, Y.; Liu, M.; Yang, Y.; Zhang, M.; Guo, Y.; Jones, G.; Liu, J.; Yin, Y. Research on the preparation of antioxidant peptides derived from egg white with assisting of high-intensity pulsed electric field. *Food Chem.* **2013**, *139*, 300–306. [CrossRef] [PubMed]

13. Lee, D.; Bamdad, F.; Khey, K.; Sunwoo, H.H. Antioxidant and anti-inflammatory properties of chicken egg vitelline membrane hydrolysates. *Poult. Sci.* **2017**, *96*, 3510–3516. [CrossRef] [PubMed]

14. Lozano-Ojalvo, D.; Molina, E.; López-Fandiño, R. Regulation of Exacerbated Immune Responses in Human Peripheral Blood Cells by Hydrolysed Egg White Proteins. *PLoS ONE* **2016**, *11*, e0151813. [CrossRef] [PubMed]

15. Gupta, D. Methods for determination of antioxidant capacity: A review. *Int. J. Pharm. Sci. Res.* **2015**, *6*, 546–566. [CrossRef]

16. Dontha, S. A review on antioxidant methods. *Asian J. Pharm. Clin. Res.* **2016**, *9*, 14–32.

17. Sheng, Z.W.; Ma, W.H.; Gao, J.H.; Bi, Y.; Zhang, W.M.; Dou, H.T.; Jin, Z.Q. Antioxidant properties of banana flower of two cultivars in China using 2,2-diphenyl-1-picrylhydrazyl (DPPH,) reducing power, 2,2′-azinobis-(3-ethylbenzthiazoline-6-sulphonate (ABTS) and inhibition of lipid peroxidation assays. *Afr. J. Biotechnol.* **2011**, *10*, 4470–4477.

18. Rodríguez-Bonilla, P.; Gandía-Herrero, F.; Matencio, A.; García-Carmona, F.; López-Nicolás, J.M. Comparative Study of the Antioxidant Capacity of Four Stilbenes Using ORAC, ABTS+, and FRAP Techniques. *Food Anal. Methods* **2017**, *164*, 191–3000. [CrossRef]

19. Ilyasov, I.; Beloborodov, V.; Selivanova, I.; Terekhov, R.P. ABTS/PP Decolorization Assay of Antioxidant Capacity Reaction Pathways. *Int. J. Mol. Sci.* **2020**, *21*, 1131. [CrossRef]

20. Wu, J. Eggs as Functional Foods and Nutraceuticals for Human Health. *Food Chem. Funct. Anal.* **2019**, *14*, 1–406.

21. Huang, X.; Tu, Z.; Xiao, H.; Wang, H.; Zhang, L.; Hu, Y.; Zhang, Q.; Niu, P. Characteristics and antioxidant activities of ovalbumin glycated with different saccharides under heat moisture treatment. *Food Res. Int.* **2012**, *48*, 866–872. [CrossRef]

22. Tu, Z.; Hu, Y.-M.; Wang, H.; Huang, X.-Q.; Xia, S.-Q.; Niu, P.-P. Microwave heating enhances antioxidant and emulsifying activities of ovalbumin glycated with glucose in solid-state. *J. Food Sci. Technol.* **2013**, *52*, 1453–1461. [CrossRef] [PubMed]

23. Yang, W.; Tu, Z.; Wang, H.; Zhang, L.; Song, Q. Glycation of ovalbumin after high-intensity ultrasound pretreatment: Effects on conformation, immunoglobulin (Ig)G/IgE binding ability and antioxidant activity. *J. Sci. Food Agric.* **2018**, *98*, 3767–3773. [CrossRef] [PubMed]

24. Nakamura, S.; Kato, A.; Kobayashi, K. Enhanced antioxidative effect of ovalbumin due to covalent binding of polysaccharides. *J. Agric. Food Chem.* **1992**, *40*, 2033–2037. [CrossRef]

25. Awatsuhara, R.; Nagao, K.; Harada, K.; Maeda, T.; Nomura, T. Antioxidative activity of the buckwheat polyphenol rutin in combination with ovalbumin. *Mol. Med. Rep.* **2009**, *3*, 121–125. [CrossRef]

26. Li, C.-P.; He, Z.; Wang, X.; Yang, L.; Yin, C.; Zhang, N.; Lin, J.; Zhao, H. Selenization of ovalbumin by dry-heating in the presence of selenite: Effect on protein structure and antioxidant activity. *Food Chem.* **2014**, *148*, 209–217. [CrossRef] [PubMed]

27. Wu, J.; Acero-Lopez, A. Ovotransferrin: Structure, bioactivities, and preparation. *Food Res. Int.* **2012**, *46*, 480–487. [CrossRef]

28. Ibrahim, H.R.; Hoq, I.; Aoki, T. Ovotransferrin possesses SOD-like superoxide anion scavenging activity that is promoted by copper and manganese binding. *Int. J. Boil. Macromol.* **2007**, *41*, 631–640. [CrossRef]

29. You, J.; Luo, Y.; Wu, J. Conjugation of Ovotransferrin with Catechin Shows Improved Antioxidant Activity. *J. Agric. Food Chem.* **2014**, *62*, 2581–2587. [CrossRef]

30. Moon, S.H.; Lee, J.H.; Ahn, D.U.; Paik, H.-D. In vitro antioxidant and mineral-chelating properties of natural and autocleaved ovotransferrin. *J. Sci. Food Agric.* **2014**, *95*, 2065–2070. [CrossRef]

31. Liu, T.; Navarro, S.; Lopata, A.L. Current advances of murine models for food allergy. *Mol. Immunol.* **2016**, *70*, 104–117. [CrossRef]

32. Sheng, L.; Su, P.; Han, K.; Chen, J.; Cao, A.; Zhang, Z.; Jin, Y.; Ma, M. Synthesis and structural characterization of lysozyme–pullulan conjugates obtained by the Maillard reaction. *Food Hydrocoll.* **2017**, *71*, 1–7. [CrossRef]

33. Hamdani, A.M.; Wani, I.A.; Bhat, N.A.; Siddiqi, R.A. Effect of guar gum conjugation on functional, antioxidant and antimicrobial activity of egg white lysozyme. *Food Chem.* **2017**, *240*, 1201–1209. [CrossRef] [PubMed]

34. Hashemi, M.M.; Aminlari, M.; Moosavinasab, M. Preparation of and studies on the functional properties and bactericidal activity of the lysozyme–xanthan gum conjugate. *LWT* **2014**, *57*, 594–602. [CrossRef]

35. Verdot, L.; Lalmanach, G.; Vercruysse, V.; Hartmann, S.; Lucius, R.; Hoebeke, J.; Gauthier, F.; Vray, B. Cystatins Up-regulate Nitric Oxide Release from Interferon-γ- activated Mouse Peritoneal Macrophages. *J. Boil. Chem.* **1996**, *271*, 28077–28081. [CrossRef] [PubMed]

36. Lehtinen, M.K.; Tegelberg, S.; Schipper, H.; Su, H.; Zukor, H.; Manninen, O.; Kopra, O.; Joensuu, T.; Hakala, P.; Bonni, A.; et al. Cystatin B deficiency sensitizes neurons to oxidative stress in progressive myoclonus epilepsy, EPM1. *J. Neurosci.* **2009**, *29*, 5910–5915. [CrossRef] [PubMed]

37. Lee, S.; Han, J.; Decker, E.A. Antioxidant Activity of Phosvitin in Phosphatidylcholine Liposomes and Meat Model Systems. *J. Food Sci.* **2002**, *67*, 37–41. [CrossRef]

38. Castellani, O.; Guérin-Dubiard, C.; David-Briand, E.; Anton, M. Influence of physicochemical conditions and technological treatments on the iron binding capacity of egg yolk phosvitin. *Food Chem.* **2004**, *85*, 569–577. [CrossRef]

39. Nimalaratne, C.; Lopes-Lutz, D.; Schieber, A.; Wu, J. Free aromatic amino acids in egg yolk show antioxidant properties. *Food Chem.* **2011**, *129*, 155–161. [CrossRef]

40. Lee, J.; Paik, H.-D. Anticancer and immunomodulatory activity of egg proteins and peptides: A review. *Poult. Sci.* **2019**, *98*, 6505–6516. [CrossRef]

41. Da Silva, A.C.; Queiroz, A.E.S.D.F.; Oliveira, J.T.C.; Medeiros, E.V.; De Souza-Motta, C.M.; Moreira, K.A. Antioxidant Activities of Chicken Egg White Hydrolysates Obtained by New Purified Protease of Aspergillus avenaceus URM 6706. *Braz. Arch. Boil. Technol.* **2019**, *62*, 14. [CrossRef]

42. Chen, G.; Zhang, X.W. Proteomics in Food Biotechnology. In *Omics Technologies: Tools for Food Science*; Benkeblia, N., Ed.; Crc Press-Taylor & Francis Group: Boca Raton, FL, USA, 2012; pp. 99–118.

43. Yuan, J.; Zheng, Y.; Wu, Y.; Chen, H.; Tong, P.; Gao, J. Double enzyme hydrolysis for producing antioxidant peptide from egg white: Optimization, evaluation, and potential allergenicity. *J. Food Biochem.* **2019**, *44*, 12. [CrossRef] [PubMed]

44. Zhang, B.; Wang, H.; Wang, Y.; Yu, Y.; Liu, J.; Liu, B.; Zhang, T. Identification of antioxidant peptides derived from egg-white protein and its protective effects on H2O2-induced cell damage. *Int. J. Food Sci. Technol.* **2019**, *54*, 2219–2227. [CrossRef]

45. Neto, Y.A.A.H.; Rosa, J.C.; Cabral, H. Peptides with antioxidant properties identified from casein, whey, and egg albumin hydrolysates generated by two novel fungal proteases. *Prep. Biochem. Biotechnol.* **2019**, *49*, 639–648. [CrossRef] [PubMed]

46. Jing, H.; Yap, M.; Wong, P.Y.Y.; Kitts, D.D. Comparison of Physicochemical and Antioxidant Properties of Egg-White Proteins and Fructose and Inulin Maillard Reaction Products. *Food Bioprocess Technol.* **2009**, *4*, 1489–1496. [CrossRef]

47. Van Der Plancken, I.; Van Loey, A.; Hendrickx, M. Combined effect of high pressure and temperature on selected properties of egg white proteins. *Innov. Food Sci. Emerg. Technol.* **2005**, *6*, 11–20. [CrossRef]

48. Karadag, A.; Özçelik, B.; Saner, S. Review of Methods to Determine Antioxidant Capacities. *Food Anal. Methods* **2009**, *2*, 41–60. [CrossRef]

49. Davalos, A.; Miguel, M.; Bartolomé, B.; López-Fandiño, R. Antioxidant Activity of Peptides Derived from Egg White Proteins by Enzymatic Hydrolysis. *J. Food Prot.* **2004**, *67*, 1939–1944. [CrossRef]

50. Ou, B.; Huang, D.; Hampsch-Woodill, M.; Flanagan, J.A.; Deemer, E.K. Analysis of Antioxidant Activities of Common Vegetables Employing Oxygen Radical Absorbance Capacity (ORAC) and Ferric Reducing Antioxidant Power (FRAP) Assays: A Comparative Study. *J. Agric. Food Chem.* **2002**, *50*, 3122–3128. [CrossRef]

51. Yu, Z.; Liu, B.; Zhao, W.; Yin, Y.; Liu, J.; Chen, F. Primary and secondary structure of novel ACE-inhibitory peptides from egg white protein. *Food Chem.* **2012**, *133*, 315–322. [CrossRef]

52. Huang, W.-Y.; Majumder, K.; Wu, J. Oxygen radical absorbance capacity of peptides from egg white protein ovotransferrin and their interaction with phytochemicals. *Food Chem.* **2010**, *123*, 635–641. [CrossRef]

53. Memarpoor-Yazdi, M.; Asoodeh, A.; Chamani, J. A novel antioxidant and antimicrobial peptide from hen egg white lysozyme hydrolysates. *J. Funct. Foods* **2012**, *4*, 278–286. [CrossRef]

54. Rao, S.; Sun, J.; Liu, Y.; Zeng, H.; Su, Y.; Yang, Y. ACE inhibitory peptides and antioxidant peptides derived from in vitro digestion hydrolysate of hen egg white lysozyme. *Food Chem.* **2012**, *135*, 1245–1252. [CrossRef] [PubMed]

55. Carrillo, W.; Gómez-Ruiz, J.A.; Miralles, B.; Ramos, M.; Barrio, D.; Recio, I. Identification of antioxidant peptides of hen egg-white lysozyme and evaluation of inhibition of lipid peroxidation and cytotoxicity in the Zebrafish model. *Eur. Food Res. Technol.* **2016**, *242*, 1777–1785. [CrossRef]

56. Liu, J.; Chen, Z.; He, J.; Zhang, Y.; Zhang, T.; Jiang, Y. Anti-oxidative and anti-apoptosis effects of egg white peptide, Trp-Asn-Trp-Ala-Asp, against H2O2-induced oxidative stress in human embryonic kidney 293 cells. *Food Funct.* **2014**, *5*, 3179–3188. [CrossRef] [PubMed]

57. Liu, J.; Jin, Y.; Lin, S.; Jones, G.S.; Chen, F. Purification and identification of novel antioxidant peptides from egg white protein and their antioxidant activities. *Food Chem.* **2015**, *175*, 258–266. [CrossRef] [PubMed]

58. Chen, C.; Chi, Y.-J.; Zhao, M.-Y.; Lv, L. Purification and identification of antioxidant peptides from egg white protein hydrolysate. *Amino Acids* **2011**, *43*, 457–466. [CrossRef]

59. Stadtman, E.R.; Moskovitz, J.; Berlett, B.S.; Levine, R.L. Cyclic oxidation and reduction of protein methionine residues is an important antioxidant mechanism. *Mol. Cell. Biochem.* **2002**, *234*, 3–9. [CrossRef]

60. Baradaran, A.; Nasri, H.; Rafieian-Kopaei, M. Oxidative stress and hypertension: Possibility of hypertension therapy with antioxidants. *J. Res. Med. Sci.* **2014**, *19*, 358–367.

61. Manso, M.A.; Miguel, M.; Even, J.; Hernández, R.; Aleixandre, A.; López-Fandiño, R. Effect of the long-term intake of an egg white hydrolysate on the oxidative status and blood lipid profile of spontaneously hypertensive rats. *Food Chem.* **2008**, *109*, 361–367. [CrossRef]

62. Garcés-Rimón, M.; Gonzalez, C.; Uranga, J.; López-Miranda, V.; López-Fandiño, R.; Miguel, M. Pepsin Egg White Hydrolysate Ameliorates Obesity-Related Oxidative Stress, Inflammation and Steatosis in Zucker Fatty Rats. *PLoS ONE* **2016**, *11*, e0151193. [CrossRef]

63. Sun, S.; Niu, H.; Yang, T.; Lin, Q.; Luo, F. Antioxidant and anti-fatigue activities of egg white peptides prepared by pepsin digestion. *J. Sci. Food Agric.* **2014**, *94*, 3195–3200. [CrossRef] [PubMed]

64. Garcés-Rimón, M.; López-Expósito, I.; López-Fandiño, R.; Miguel, M. Egg white hydrolysates with in vitro biological multiactivities to control complications associated with the metabolic syndrome. *Eur. Food Res. Technol.* **2015**, *242*, 61–69. [CrossRef]

65. Xiang, J.Z.; Hong, Y.W.; Cai, Y.Y.; Qing, S.M.; Ji, Z. Technology optimization, antioxidant activities and characteristics of peptide from egg white. *J. Food Sci. Biotechnol.* **2013**, *32*, 844–853.

66. Lin, S.Y.; Guo, Y.; Liu, J.B.; You, Q.; Yin, Y.G.; Cheng, S. Optimized enzymatic hydrolysis and pulsed electric field treatment for production of antioxidant peptides from egg white protein. *Afr. J. Biotechnol.* **2011**, *10*, 11648–11657.

67. Cho, D.-Y.; Jo, K.; Cho, S.Y.; Kim, J.M.; Lim, K.; Suh, H.J.; Oh, S. Antioxidant Effect and Functional Properties of Hydrolysates Derived from Egg-White Protein. *Food Sci. Anim. Resour.* **2014**, *34*, 362–371. [CrossRef]

68. Noh, D.O.; Suh, H.J. Preparation of Egg White Liquid Hydrolysate (ELH) and Its Radical-Scavenging Activity. *Prev. Nutr. Food Sci.* **2015**, *20*, 183–189. [CrossRef]

69. Tanasković, S.J.; Luković, N.; Grbavčić, S.; Stefanović, A.; Jovanovic, J.; Bugarski, B.; Jugovic, Z.K. Production of egg white protein hydrolysates with improved antioxidant capacity in a continuous enzymatic membrane reactor: Optimization of operating parameters by statistical design. *J. Food Sci. Technol.* **2017**, *55*, 128–137. [CrossRef]

70. Hernández-Ledesma, B.; Miralles, B.; Amigo, L.; Ramos, M.; Recio, I. Identification of antioxidant and ACE-inhibitory peptides in fermented milk. *J. Sci. Food Agric.* **2005**, *85*, 1041–1048. [CrossRef]

71. Singh, A.; Ramaswamy, H.S. Effect of high-pressure treatment on trypsin hydrolysis and antioxidant activity of egg white proteins. *Int. J. Food Sci. Technol.* **2013**, *49*, 269–279. [CrossRef]

72. Stefanović, A.B.; Jovanovic, J.; Grbavčić, S.Ž.; Šekuljica, N.Ž.; Manojlović, V.B.; Bugarski, B.M.; Knežević-Jugović, Z.D. Impact of ultrasound on egg white proteins as a pretreatment for functional hydrolysates production. *Eur. Food Res. Technol.* **2014**, *239*, 979–993. [CrossRef]

73. Remanan, M.K.; Wu, J. Antioxidant activity in cooked and simulated digested eggs. *Food Funct.* **2014**, *5*, 1464–1474. [CrossRef] [PubMed]

74. Wang, J.; Liao, W.; Nimalaratne, C.; Chakrabarti, S.; Wu, J. Purification and characterization of antioxidant peptides from cooked eggs using a dynamic in vitro gastrointestinal model in vascular smooth muscle A7r5 cells. *NPJ Sci. Food* **2018**, *2*, 7. [CrossRef] [PubMed]

75. Chakrabarti, S.; Guha, S.; Majumder, K. Food-Derived Bioactive Peptides in Human Health: Challenges and Opportunities. *Nutrients* **2018**, *10*, 1738. [CrossRef] [PubMed]

76. Jahandideh, F.; Majumder, K.; Chakrabarti, S.; Morton, J.S.; Panahi, S.; Kaufman, S.; Davidge, S.T.; Wu, J. Beneficial Effects of Simulated Gastro-Intestinal Digests of Fried Egg and Its Fractions on Blood Pressure, Plasma Lipids and Oxidative Stress in Spontaneously Hypertensive Rats. *PLoS ONE* **2014**, *9*, e115006. [CrossRef] [PubMed]

77. Sakanaka, S.; Tachibana, Y.; Ishihara, N.; Juneja, L.R. Antioxidant activity of egg-yolk protein hydrolysates in a linoleic acid oxidation system. *Food Chem.* **2004**, *86*, 99–103. [CrossRef]

78. Zambrowicz, A.; Pokora, M.; Eckert, E.; Szołtysik, M.; Dąbrowska, A.; Chrzanowska, J.; Trziszka, T. Antioxidant and antimicrobial activity of lecithin free egg yolk protein preparation hydrolysates obtained with digestive enzymes. *Funct. Foods Health Dis.* **2012**, *2*, 487. [CrossRef]

79. Pokora, M.; Eckert, E.; Zambrowicz, A.; Bobak, L.; Szołtysik, M.; Dąbrowska, A.; Chrzanowska, J.; Polanowski, A.; Trziszka, T. Biological and functional properties of proteolytic enzyme-modified egg protein by-products. *Food Sci. Nutr.* **2013**, *1*, 184–195. [CrossRef]

80. Ishikawa, S.-I.; Yano, Y.; Arihara, K.; Itoh, M. Egg Yolk Phosvitin Inhibits Hydroxyl Radical Formation from the Fenton Reaction. *Biosci. Biotechnol. Biochem.* **2004**, *68*, 1324–1331. [CrossRef]

81. Yousr, M.N.; Howell, N.K. Antioxidant and ACE Inhibitory Bioactive Peptides Purified from Egg Yolk Proteins. *Int. J. Mol. Sci.* **2015**, *16*, 29161–29178. [CrossRef]

82. Zambrowicz, A.; Pokora, M.; Setner, B.; Dąbrowska, A.; Szołtysik, M.; Babij, K.; Szewczuk, Z.; Trziszka, T.; Lubec, G.; Chrzanowska, J. Multifunctional peptides derived from an egg yolk protein hydrolysate: Isolation and characterization. *Amino Acids* **2014**, *47*, 369–380. [CrossRef]

83. Yoo, H.; Bamdad, F.; Gujral, N.; Suh, J.-W.; Sunwoo, H. High Hydrostatic Pressure-Assisted Enzymatic Treatment Improves Antioxidant and Anti-inflammatory Properties of Phosvitin. *Curr. Pharm. Biotechnol.* **2017**, *18*, 158–167. [CrossRef] [PubMed]

84. Jiang, B.; Mine, Y. Preparation of novel functional oligophosphopeptides from hen egg yolk phosvitin. *J. Agric. Food Chem.* **2000**, *48*, 990–994. [CrossRef] [PubMed]

85. Feng, F.; Mine, Y. Phosvitin phosphopeptides increase iron uptake in a Caco-2 cell monolayer model. *Int. J. Food Sci. Technol.* **2006**, *41*, 455–458. [CrossRef]

86. Katayama, S.; Xu, X.; Fan, M.Z.; Mine, Y. Antioxidative Stress Activity of Oligophosphopeptides Derived from Hen Egg Yolk Phosvitin in Caco-2 Cells. *J. Agric. Food Chem.* **2006**, *54*, 773–778. [CrossRef]

87. Xu, X.; Katayama, S.; Mine, Y. Antioxidant activity of tryptic digests of hen egg yolk phosvitin. *J. Sci. Food Agric.* **2007**, *87*, 2604–2608. [CrossRef]

88. Young, D.; Nau, F.; Pasco, M.; Mine, Y. Identification of Hen Egg Yolk-Derived Phosvitin Phosphopeptides and Their Effects on Gene Expression Profiling against Oxidative Stress-Induced Caco-2 Cells. *J. Agric. Food Chem.* **2011**, *59*, 9207–9218. [CrossRef]

89. Young, D.; Fan, M.Z.; Mine, Y. Egg Yolk Peptides Up-regulate Glutathione Synthesis and Antioxidant Enzyme Activities in a Porcine Model of Intestinal Oxidative Stress. *J. Agric. Food Chem.* **2010**, *58*, 7624–7633. [CrossRef]

90. Park, P.-J.; Jung, W.-K.; Nam, K.-S.; Shahidi, F.; Kim, S.-K. Purification and characterization of antioxidative peptides from protein hydrolysate of lecithin-free egg yolk. *J. Am. Oil Chem. Soc.* **2001**, *78*, 651–656. [CrossRef]

91. Zambrowicz, A.; Eckert, E.; Pokora, M.; Dabrowska, A.; Szoltysik, M.; Bobak, L.; Trziszka, T.; Chrzanowska, J. Biological activity of egg-yolk protein by-product hydrolysates obtained with the use of non-commercial plant protease. *Ital. J. Food Sci.* **2015**, *27*, 450–458.

92. Eckert, E.; Zambrowicz, A.; Bobak, L.; Zabłocka, A.; Chrzanowska, J.; Trziszka, T. Production and Identification of Biologically Active Peptides Derived from By-product of Hen Egg-Yolk Phospholipid Extraction. *Int. J. Pept. Res. Ther.* **2018**, *25*, 669–680. [CrossRef]

93. Sakanaka, S.; Tachibana, Y. Active oxygen scavenging activity of egg-yolk protein hydrolysates and their effects on lipid oxidation in beef and tuna homogenates. *Food Chem.* **2006**, *95*, 243–249. [CrossRef]

94. Eckert, E.; Zambrowicz, A.; Pokora, M.; Dabrowska, A.; Szoltysik, M.; Chrzanowska, J.; Trziszka, T. application of microbial proteases to obtain egg yolk protein hydrolysates with antioxidant and antimicrobial activity. *ZYWN Nauk Technol. Jakosc* **2013**, *20*, 105–118. [CrossRef]

95. Jung, S.; Jo, C.; Kang, M.-G.; Ahn, D.U.; Nam, K.-C. Elucidation of Antioxidant Activity of Phosvitin Extracted from Egg Yolk using Ground Meat. *Food Sci. Anim. Resour.* **2012**, *32*, 162–167. [CrossRef]

96. Shi, Y.; Kovacs-Nolan, J.; Jiang, B.; Tsao, R.; Mine, Y. Antioxidant activity of enzymatic hydrolysates from eggshell membrane proteins and its protective capacity in human intestinal epithelial Caco-2 cells. *J. Funct. Foods* **2014**, *10*, 35–45. [CrossRef]

97. Shi, Y.; Kovacs-Nolan, J.; Jiang, B.; Tsao, R.; Mine, Y. Peptides derived from eggshell membrane improve antioxidant enzyme activity and glutathione synthesis against oxidative damage in Caco-2 cells. *J. Funct. Foods* **2014**, *11*, 571–580. [CrossRef]

98. Nimalaratne, C.; Wu, J. Hen Egg as an Antioxidant Food Commodity: A Review. *Nutrients* **2015**, *7*, 8274–8293. [CrossRef]

99. Jahandideh, F.; Chakrabarti, S.; Davidge, S.T.; Wu, J. Antioxidant Peptides Identified from Ovotransferrin by the ORAC Method Did Not Show Anti-Inflammatory and Antioxidant Activities in Endothelial Cells. *J. Agric. Food Chem.* **2015**, *64*, 113–119. [CrossRef]

100. You, S.-J.; Udenigwe, C.C.; E Aluko, R.; Wu, J. Multifunctional peptides from egg white lysozyme. *Food Res. Int.* **2010**, *43*, 848–855. [CrossRef]

101. Xu, M.; Shangguan, X.; Wang, W.; Chen, J. Antioxidative activity of hen egg ovalbumin hydrolysates. *Asia Pac. J. Clin. Nutr.* **2007**, *16*, 178–182.

102. Abeyrathne, E.D.N.S.; Lee, H.Y.; Jo, C.; Nam, K.C.; Ahn, D.U. Enzymatic hydrolysis of ovalbumin and the functional properties of the hydrolysates. *Poult. Sci.* **2014**, *93*, 2678–2686. [CrossRef]

103. Abeyrathne, E.D.N.S.; Lee, H.Y.; Jo, C.; Suh, J.W.; Ahn, D.U. Enzymatic hydrolysis of ovomucoid and the functional properties of its hydrolysates. *Poult. Sci.* **2015**, *94*, 2280–2287. [CrossRef] [PubMed]

104. Abeyrathne, E.; Lee, H.; Jo, C.; Suh, J.; Ahn, D.U. Enzymatic hydrolysis of ovomucin and the functional and structural characteristics of peptides in the hydrolysates. *Food Chem.* **2016**, *192*, 107–113. [CrossRef] [PubMed]

105. Chang, O.K.; Ha, G.E.; Han, G.-S.; Seol, K.-H.; Kim, H.W.; Jeong, S.-G.; Oh, M.H.; Park, B.-Y.; Ham, J.-S. Novel Antioxidant Peptide Derived from the Ultrafiltrate of Ovomucin Hydrolysate. *J. Agric. Food Chem.* **2013**, *61*, 7294–7300. [CrossRef] [PubMed]

MDPI

St. Alban-Anlage 66

4052 Basel

Switzerland

Tel. +41 61 683 77 34

Fax +41 61 302 89 18

www.mdpi.com

Foods Editorial Office

E-mail: foods@mdpi.com

www.mdpi.com/journal/foods

9 783036 525204